Molecular Biology Intelligence Unit

Molecular Aspects of Spermatogenesis

Dominic Poccia, Ph. D.

Amherst College
Amherst, Massachusetts

R.G. Landes Company
Austin

MOLECULAR BIOLOGY INTELLIGENCE UNIT
MOLECULAR ASPECTS OF SPERMATOGENESIS
R.G. LANDES COMPANY
Austin

CRC Press is the exclusive worldwide distributor of publications
of the Molecular Biology Intelligence Unit.
CRC Press, 2000 Corporate Blvd., NW, Boca Raton, FL 33431. Phone: 407/994-0555.

Submitted: June 1994
Published: September 1994

Please address all inquiries to the Publisher:
R.G. Landes Company, 909 Pine Street, Georgetown, TX 78626
or
P.O. Box 4858, Austin, TX 78765

Phone: 512/ 863 7762; FAX: 512/ 863 0081

ISBN 1-57059-204-7
CATALOG # LN9204

Library of Congress Cataloging-in-Publication Data

Poccia, Dominic
Molecular aspects of spermatogenesis / Dominic Poccia.
p. cm.—(Molecular biology intelligence unit)
Includes bibliographical references and index.
ISBN 1-57059-204-7
1. Spermatogenesis in animals. 2. Reproduction—Molecular aspects.
I. Title. II. Series.
QL966.P55 1994 94-31867
591.3'2—dc20 CIP

While the authors, editors and publisher believe that drug selection and dosage and the specifications and usage of equipment and devices, as set forth in this book, are in accord with current recommendations and practice at the time of publication, they make no warranty, expressed or implied, with respect to material described in this book. In view of the ongoing research, equipment development, changes in governmental regulations and the rapid accumulation of information relating to the biomedical sciences, the reader is urged to carefully review and evaluate the information provided herein.

To Our Readers

R.G. Landes Company publishes four book series: *Medical Intelligence Unit, Molecular Biology Intelligence Unit, Neuroscience Intelligence Unit,* and *Biotechnology Intelligence Unit.* Our goal is to publish the most recent information in biomedical science for sophisticated researchers and physicians.

To achieve this goal we have accelerated our publishing program to conform to the fast pace in which information grows in biomedical science. The book you have in hand, like all titles in our series, was published *within 90 days of receipt of the manuscript.*

As you might expect, this causes a few problems for us; sometimes it makes our office more like a big city newspaper than a scholarly publisher. So as you look through this book you may see something that isn't just right. Please let us know. Or if you have an idea for improving our books, we'd like very much to hear from you. If the problem you describe hasn't already been discovered or if your idea is provocative enough to promote a discussion here, we'll give you a free book of your choice. Just list three titles in order of preference and we'll send you one, based on availability—with our thanks. Our address is printed on the copyright page of each of our books.

Elinda McKenna
Director of Operations
R.G. Landes Company

Dedication

This book is dedicated to my parents, Louis and Frances Poccia, who continually encouraged me by providing large doses of independence and an absolute unwavering support of my activities, while never hesitating to provide the clearest human guidelines in matters of importance.

CONTENTS

PREFACE

The study of spermatogenesis comprises a vast field in which reproductive biologists, morphologists, geneticists, biochemists, molecular biologists, endocrinologists and physicians find common interests. Most research efforts have emphasized mammalian systems, reflecting our generally anthropomorphic outlook, concerns about problems of fertility and contraception, and, as always, funding opportunities. Nonetheless, major contributions have been made by the study of various invertebrates and non-mammalian vertebrates, especially organisms which provide a combination of genetic and biochemical approaches. Although the emphasis in this review is overwhelmingly mammalian, and mostly rodent, examples from fruit flies, worms, sea urchins, frogs and other creatures are included when appropriate.

While concerns about population control, sterility or disease may fuel much research on spermatogenesis, it is a biological field intellectually fascinating in its own right. Such fundamental cellular and developmental processes as meiosis, stem cell renewal, determination, extreme cytodifferentiation, short and long range cell-cell signaling, and cell-specific gene expression are all crucial to the success of the spermatogenic program. This program terminates with a usually motile cell carrying a nuclear payload capable of contributing half the genetic complement of a newly formed individual at fertilization. There is clearly much room for researchers of a wide range of talents and interests in the field of spermatogenesis.

Work in spermatogenesis is already immense and still growing. As noted by Yves Clermont, the symposium on the Biology of the Testes organized by the New York Academy of Sciences in 1952 had 16 participants; a 1990 meeting sponsored by the same organization had 350 attendees.[1] The meeting of the 6th International Congress on Spermatology in Siena, Italy in 1990 was attended by 300 delegates from 20 countries and produced 193 presentations. The BIOSIS (c) database lists 132 papers indexed to "spermatogenesis" in 1970 and 443 in 1993. The number of papers indexed to "sperm" increased 350% in the last 20 years to 1814 references for 1993.

---

1. Clermont, Y. Four decades of research on the biology of the male reproductive system: A few landmarks. Ann. NY Acad. Sci. 1991; 637:17-25.

Clearly, the field of spermatogenesis is too vast to provide a comprehensive review in these pages. Instead, I have emphasized work of the last two years, discussing earlier observations only for clarity or perspective. I trust that the papers referenced will lead the interested reader quickly to the important papers of the older literature. References have also been given to many excellent and thorough recent reviews in the appropriate sections.

Given the diversity of the potential audience with an interest in recent progress in molecular aspects of spermatogenesis, I have tried to provide an entree into the literature with at least a minimal background and rationale for approaches taken, sometimes at the risk of simplifying for a sophisticated reader information that might be of use to a novice in a particular subfield.

Dominic Poccia, PhD
June, 1994

Acknowledgments

I would like to thank those who have kindled my interest in spermatogenesis through their writings, collaborated with me in spermatogenesis research, or made possible the time and resources for doing so: Dr. David Nishioka, Dr. Ray Green, Dr. Geoffrey Childs, Dr. Fred Wilt, Dr. Steve Benson, Dr. Michael Risley, Dr. David P. Bloch, Dr. Daniel Mazia, Dr. Marcia Simpson, Dr. Connie Kostacos, Dr. Laura Goetzl, Dr. David McDermott, Ms. Jenny Sue Smith, Ms. Monica Jones, Mr. Stephen Tung, the National Institutes of Health, Amherst College, and the University of Massachusetts, Amherst. Special thanks to my son, Joseph, for his pen and ink illustrations.

CHAPTER 1

INTRODUCTION

Spermatogenesis takes place in a unique environment created by the testis. The mammalian testis consists of an interstitial compartment containing Leydig cells, macrophages, blood vessels, lymphatic vessels, and fibroblast-like cells surrounding the seminiferous tubules which are delimited by peritubular myoid cells and contain Sertoli cells and germ cells. The Sertoli cells are connected to each other and to the germ cells by various specializations. The germ cells consist of mitotically cycling spermatogonia including a stem cell population, spermatocytes undergoing meiosis, and differentiating haploid spermatids which lead to the mature spermatozoa.

As a prelude to discussion of the molecules of spermatogenesis, this chapter introduces the main somatic and germ line cells of the testis, describes the structure of the vertebrate testis, and considers developmental changes in cell populations. The major structural features of the mammalian testis can be seen in Fig. 1.1.[1]

INTERSTITIAL COMPARTMENT

The interstitium contains undifferentiated cells during fetal and adult life. Fibroblasts produce type I and type III collagen, macrophages contribute phagocytic activity, and Leydig cells make testosterone and other important molecules.[2] Blood vessels are lined with endothelial cells and vascular smooth muscle.

The cells which have received the most attention are the Leydig cells which in the adult are polygonal. They are sometimes connected by gap junctions and contain lipid droplets and well developed smooth and rough endoplasmic reticulum. Leydig cells appear in two or more generations: fetal, adult and sometimes perinatal.[3,4] Adult Leydig cells probably arise from undifferentiated spindle shaped interstitial cell precursors in the mesenchyme.[5] Leydig cells are produced in two bursts in the rat: one between 14-28 days postpartum and one between 28-56 days.

The relative numbers of interstitial cells of the testis change with postnatal development. In CBA mice, Leydig cells increase between 11 and 31 days after birth, then decline slightly to 9×10^5 per testis.[6] Mesenchymal cells remain constant up to day 36 and decrease slightly thereafter, becoming only 4% of the total interstitial population due to increases in the other cell types. Macrophages remain almost constant at

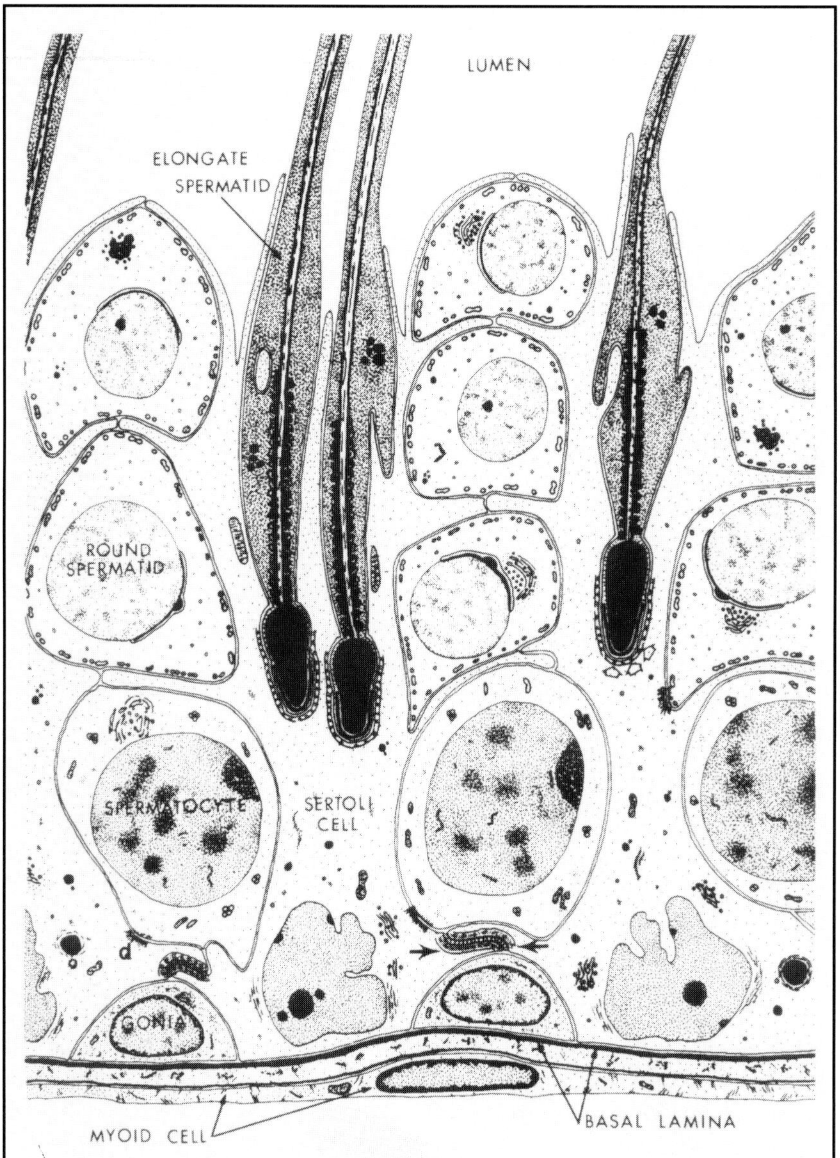

Fig. 1.1. Section of vertebrate testis. Paired arrows indicate occluding junctions. Open arrows indicate ectoplasmic specializations. Desmosome-like junctions are indicated by d. Reproduced with permission from Russell et al. Histological and histopathological evaluation of the testis. 1990[1] ©Cache River Press

1-2.5 x 10^5. Myoid, lymphatic and endothelial cells increase from the third to the fifth postnatal weeks to 6.5 x 10^5, 3.7 x 10^5 and 4.5 x 10^5 respectively. Perivascular cells remain constant at 2 x 10^5.

Leydig cell precursors in humans can be detected in the eight week fetus and by the 14th week Leydig cells increase to a maximal 24 x 10^6 per testis.[7] In the rat, fetal Leydig cells are seen at 15-17 days and reach a maximum of 130 x 10^3 at 19.5 days.[8]

SEMINIFEROUS TUBULES

Mammalian spermatogenesis takes place in the seminiferous tubules. Within the tubules, somatic *Sertoli cells* create two regions: a basal compartment containing early male germ cells and open directly to the blood vessels and an adluminal compartment containing the more advanced germ cells separated from the blood vessels by a permeability barrier.

MALE GERM CELL TYPES

The earliest germ line cells are *primordial germ cells*, a totipotent population of germ cell precursors. These exist in small numbers and little is known about them, posing questions about how they are determined, and what their patterns of gene expression are. Primordial germ cells in the mouse have been identified by surface antigens as well as the more common alkaline phosphatase activity.[9] A monoclonal antibody to one of the surface antigens has been used to isolate primordial germ cells by flow cytometry.[10] Primordial germ cells migrate from the yolk sac endoderm to the gonadal ridge, the site of the developing gonad where they eventually give rise to *spermatogonia*.

Several spermatogonial subtypes can be distinguished: the non-cycling A_0 cells that resist irradiation and repopulate the gonad following irradiation[11] and the $A_{1\text{-}4}$, intermediate and Type B spermatogonia that cycle and give rise to meiotic cells.[12] In rodents, the Type A cells have pale nuclei, the type B darker heterochromatic nuclei.[13] In primates two kinds of Type A spermatogonia have been distinguished: A-pale and A-dark, the latter serving as a reserve population and the former as an active stem cell population with the intermediate type lacking.[14]

The diploid Type B cell that commits to meiosis becomes a *primary spermatocyte*. The premeiotic spermatocyte replicates its DNA to give a cell with four times the haploid amount of nuclear DNA. Several stages of prophase I of meiosis can be distinguished by chromosomal configuration: leptotene, zygotene, pachytene and diplotene. Homologous chromosome pairing and crossing-over take place during prophase I aided by the formation of a *synaptonemal complex* between paired chromosomes. Two rounds of cell division follow without further nuclear replication. Thus a single tetraploid primary spermatocyte through its two diploid *secondary spermatocyte* progeny may yield four haploid *spermatids*.

Extreme cytodifferentiation of the early spermatids converts them through a variable number of stages to spermatozoa. Major cytological changes include extreme condensation of the chromatin, development of a flagellum for motility and an acrosome for degradation of egg vestments, coalescence of mitochondria to the midpiece, and shedding of excess cytoplasm in the form of *residual bodies*. This differentiation phase is called *spermiogenesis* and several stages can be distinguished (19 in the rat, 16 in the mouse) by the presence and placement of the Golgi apparatus and acrosome, the length and position of the flagellum and the size, shape and degree of chromatin condensation of the nucleus (Fig. 1.2).[15]

The resulting *spermatozoon* is highly specialized and unmistakable. Most aquatic metazoa produce a "primitive" type sperm with a compact head with an acrosome and nucleus, a midpiece containing one or a few mitochondria and a long flagellum or tail. Most internally fertilizing species produce sperm with more elaborate modifications but which are still recognizable by the basic subdivisions seen in the primitive sperm.[16] Many unusual forms, such as non-motile or non-flagellated sperms are produced by invertebrates. A comparative view of the ultrastructure of spermatogenesis in various animals can be found in ref. 17.

Seminiferous Cycle

The seminiferous tubules consist of an outer layer of *peritubular myoid cells* separated by a basement lamina from the seminiferous epithelium. The *seminiferous epithelium* consists of *Sertoli cells* which stop proliferating after puberty in most mammals, and the developing male germ cells. The germ cell population continually renews due to stem cells which upon division produce one daughter cell identical to the mother cell, thus maintaining a reserve of stem cells. Well-synchronized development of the germ cell cohort connected by cytoplasmic bridges gives rise to associations of groups of germ cells at various points along the length of

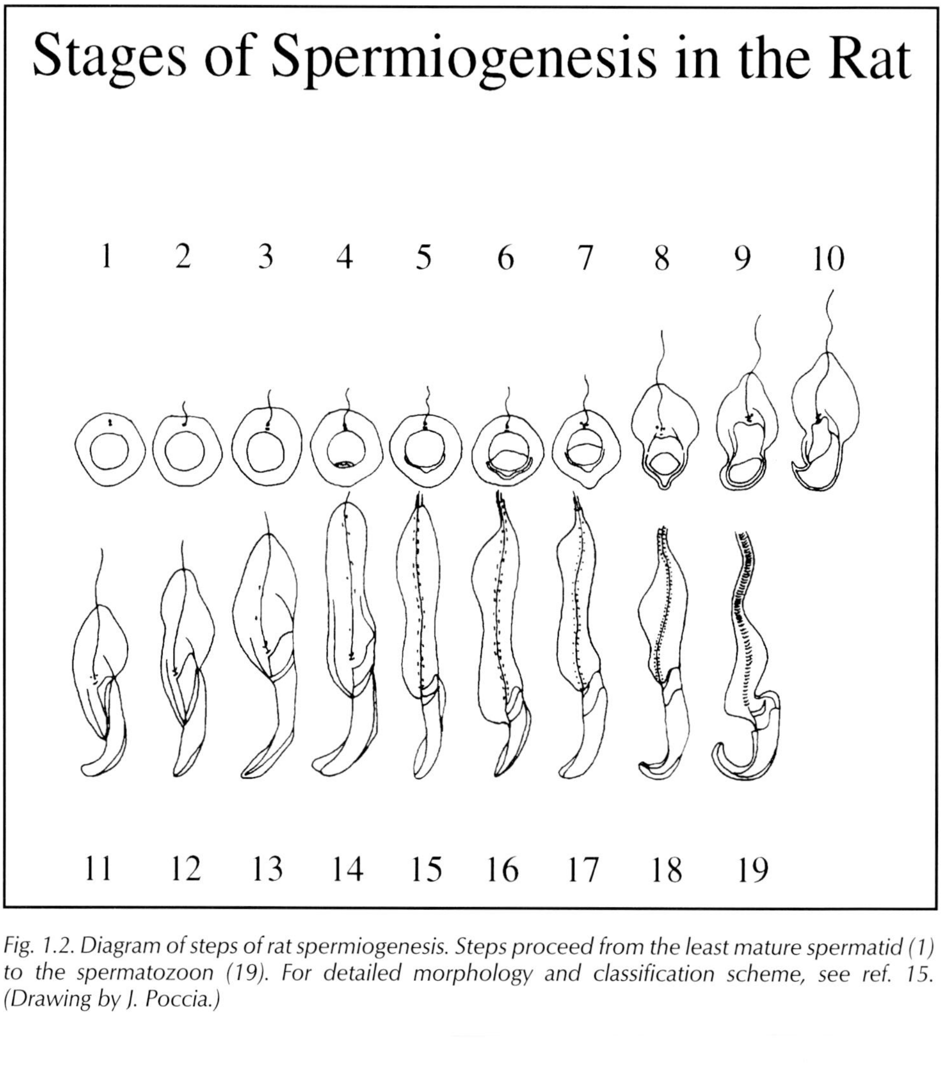

Fig. 1.2. Diagram of steps of rat spermiogenesis. Steps proceed from the least mature spermatid (1) to the spermatozoon (19). For detailed morphology and classification scheme, see ref. 15. (Drawing by J. Poccia.)

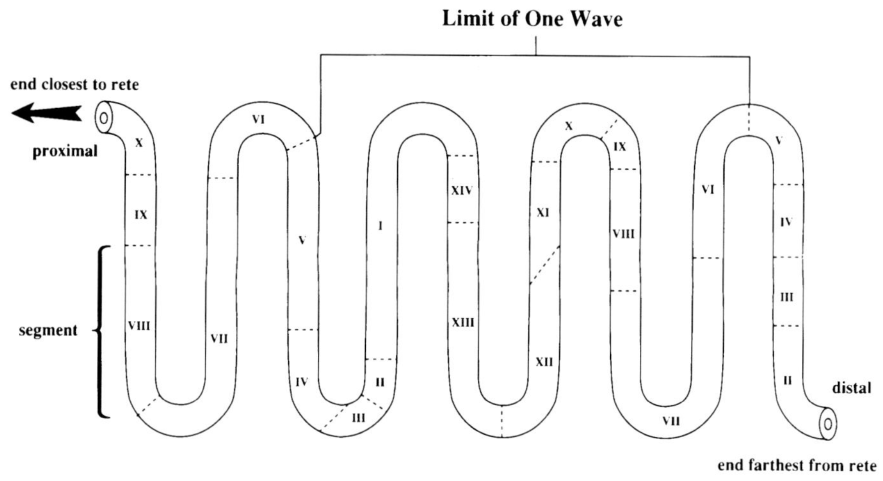

Fig. 1.3. Diagram of a longitudinal section of a seminiferous tubule from the rat, indicating the extent of a spermatogenic wave. Reproduced with permission from Russell et al. Histological and histopathological evaluation of the testis. 1990[1] ©Cache River Press

the tubule (Fig. 1.3).[1] Successive cross-sections of the tubule eventually lead back to the initial configuration or *stage* (Figs. 1.4 and 1.5). The succession of stages constitutes the seminiferous epithelial cycle (14 in the rat,[18] 12 in the mouse, 6 in man[19]), the group of stages making up the wave of the seminiferous epithelium.[20,21]

Germ Cell Kinetics

In mouse, spermatogonial stem cells, A_{single} or A_s, proliferate to produce more of the same or two paired A_{pr} cells connected by an intercellular bridge.[22] A_{pr} cells can divide further to form nests of $A_{aligned}$ or A_{al} cells. All these cells are undifferentiated spermatogonia. Almost all A_{al} cells differentiate to A_1 type spermatogonia which after six divisions give rise to spermatocytes.

A recent quantitative study of testicular whole mounts in C3H/101F_1 hybrid mice indicates that undifferentiated spermatogonia proliferate from Stage XI to IV. Within each epithelial cycle, three divisions of undifferentiated spermatogonia result in A_1 and more undifferentiated spermatogonia. At stage VIII, about 2/3 of A_{pr} and all A_{al} spermatogonia become A_1 spermatogonia.[23] There are approximately 3.3 x 10^5 undifferentiated spermatogonia and 2.5 x 10^6 differentiating spermatogonia per testis with 3.5 x 10^4 stem cells. The stem cells account for about 10% of undifferentiated spermatogonia and a little over 1% of all spermatogonia. The number of spermatocytes was estimated at about 25 x 10^6 and spermatids at about 10^7 assuming no cellular degeneration, so the actual numbers produced are likely to be somewhat higher. Some spermatogonia degenerate, apparently by apoptosis.[24] In CBA mice, Type A spermatogonia increase from 0.2 x 10^5 to 3 x 10^5 at day 21 *postpartum* and then remain constant.[6]

A spermatogenic cycle takes 12.5 days in the rat,[13] 16 days in man,[25] 9-12 days in non-human primates.[26] The duration of spermatogenesis

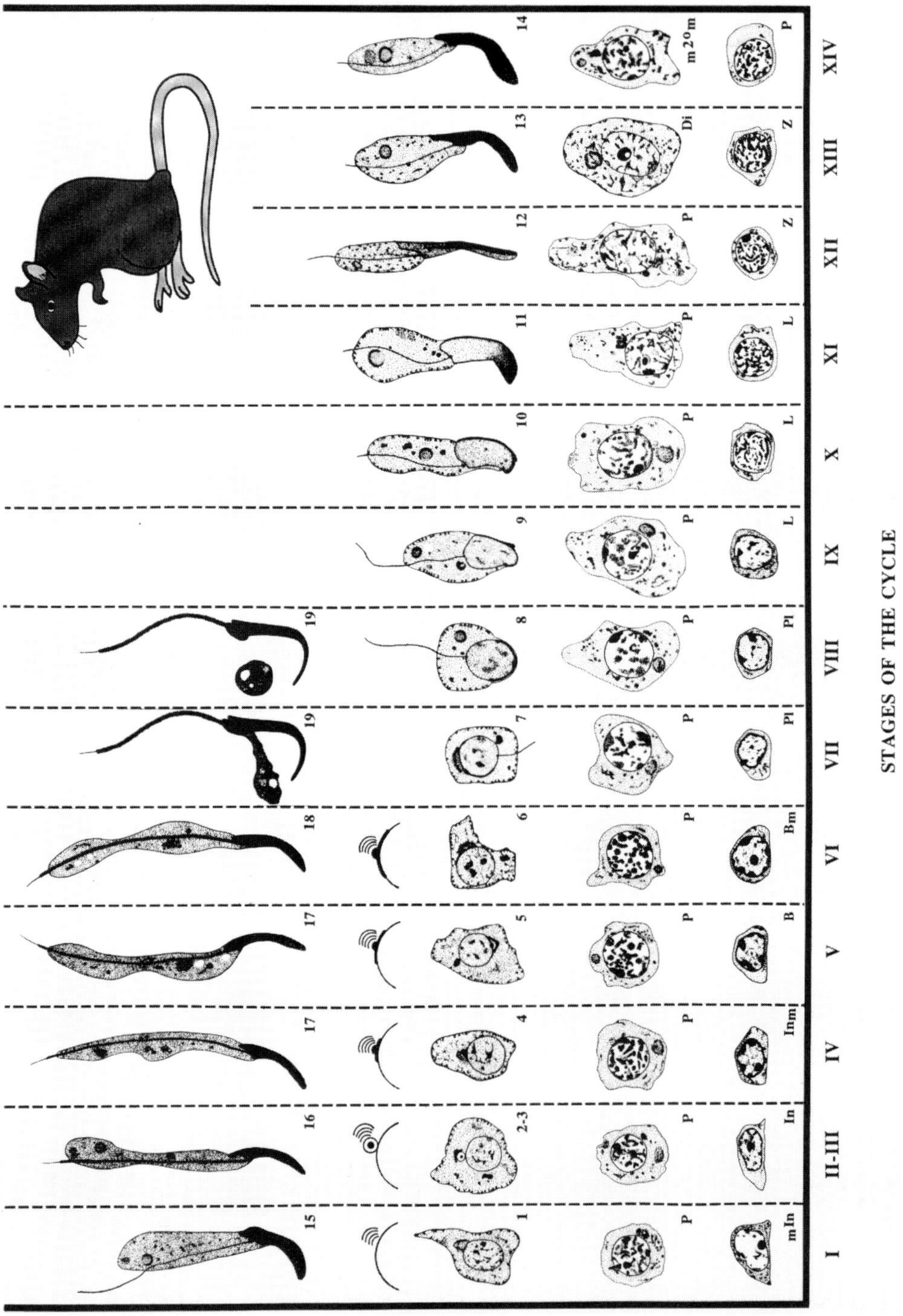

Fig. 1.4. Stages of the seminiferous cycle of the rat. Reproduced with permission from Russell et al. Histological and histopathological evaluation of the testis. 1990[1] ©Cache River Press

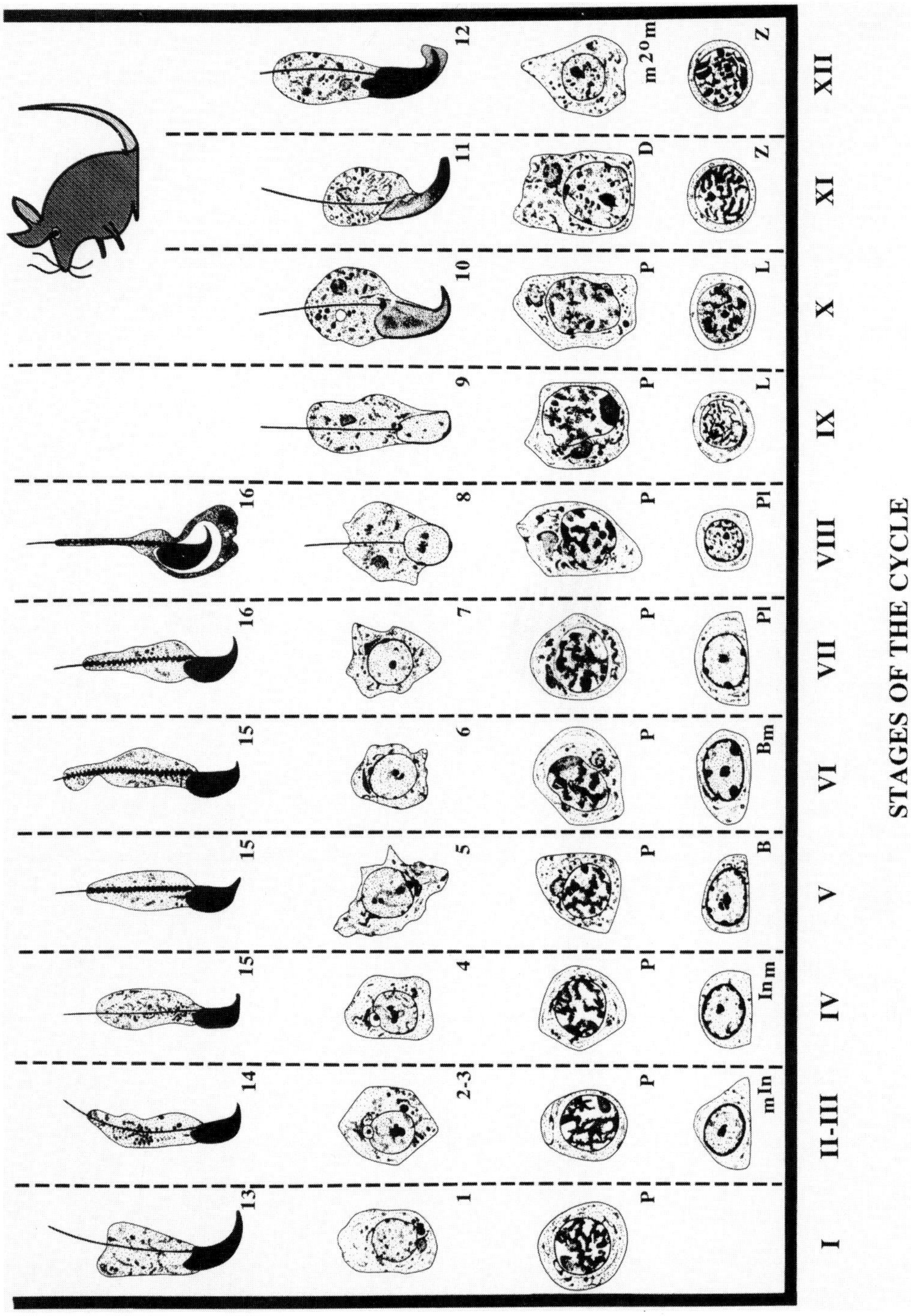

Fig. 1.5. Stages of the seminiferous cycle of the mouse. Reproduced with permission from Russell et al Histological and histopathological evaluation of the testis. 1990[1] ©Cache River Press

from spermatogonium to spermatozoon is about 35 days in mice,[27] 50 days in rats, 75 days in humans and 40 days in non-human primates.[28] The same period is about 19 days in the fruit fly *Drosophila hydei* and 10 days in *Drosophila melanogaster*,[29] about 14 days in the sea urchin *S. purpuratus*,[30] 28 days in the frog *Xenopus laevis*[31] and only 7 hours in the nematode *C. elegans*.[32]

Sertoli Cells

Perhaps more is known about Sertoli cells than any other somatic cells of the testis. These complex and multifaceted cells help create the microenvironment in which the male germ cells develop. They are intimately associated with the germ cells, and modify their extracellular milieu. They have been considered in several thorough reviews[2,20,33-36] and are the subject of a recent book.[37]

The large cells (70-80 μm long) were first described by Sertoli in 1865,[38] received little attention until the 1950s[39] and are now the subject of very active investigation. At 14.5 days in rat, Sertoli cell precursors can be detected and begin to organize structures with primordial germ cells which have migrated to the gonad. By 16.5 days aggregates of germ cells and Sertoli cells fuse to create testicular cords.[40] Sertoli cells proliferate during prepubertal development. Sertoli cells in CBA mice increase during the first three weeks to the adult level of 1.8×10^6. The number of Sertoli cells per adult testis in $C3H/101F_1$ mice is 3×10^6.[23] After this, no further Sertoli cell division is seen.[2,6]

The Sertoli cell is important in establishing a microenvironment for the developing male germ cells. During puberty, the lumen of the seminiferous tubule appears and Sertoli-Sertoli junctions are established. A network of tight junctional complexes between adjacent Sertoli cells forms the basis of the separation of the two compartments of the germinal layer: the adluminal and the basal (Fig. 1.1).[41] The occluding nature of the junctions is shown by exclusion of La^{++}.[42] Spermatogonia and preleptotene spermatocytes are confined to the basal compartment, which is accessible to the blood. The more advanced spermatocytes, spermatids and spermatozoa are located in the adluminal compartment. Therefore apart from germ cell components, the adluminal compartment should contain only those substances made or transported by the Sertoli cells. In fact, the composition of the tubule lumen is very different from that of interstitial fluid, blood or lymph.[43] The presence of compartments allows bidirectional Sertoli secretion.[44] The Sertoli tight junctions form the major part of the blood-testis barrier, but the peritubular layer of myoid cells also displays junctions which in some species (or in some regions of the tubules) may help in establishing the blood-testis barrier.[41] The tight junctions of Sertoli cells develop at 16-19 days in the rat.[45] Their maintenance may require testosterone.[46] Sertoli cells are also connected to the germ cells by gap junctions[47] and desmosomes.[48] Their associated microfilaments are disrupted by cadmium which compromises the barrier in vivo.[49] The intimate association of the Sertoli cells with developing germ cells and the formation of a special microenvironment by junctional complexes underscores the important role the Sertoli cells play in spermatogenesis.

Surface specializations between Sertoli cells and germ cells are both intimate and complex. Sertoli-germ cell specializations include spermatogonial processes, desmosome-like junctions, gap junctions, ectoplasmic specializations, spermatid processes, Sertoli cell processes, and tubulobulbar complexes.[20] *Spermatogonial processes* are projections of Type B and intermediate spermatogonial cytoplasms into the Sertoli cell made of tubular and bulbous portions. They are presumed to be involved in transfer of substances and for cell anchoring. *Desmosome-like junctions* are often found in tight junctional regions and especially between Sertoli cells and spermatogonia and between Sertoli cells and pachytene spermatocytes and serve as adhering junctions. *Gap junctions* between Sertoli cells and spermatogonia, spermatocytes and early spermatids allow passage of small molecules through their cytoplasms. *Ectoplasmic specializations* within the Sertoli cells, rich in microfilaments, are involved in attachment to spermatocytes and early spermatids and may also be involved in the transport of spermatids within the germinal epithelium and in shaping of sperm heads. *Spermatid processes* from spermatids and Sertoli cells into spermatids may function in transfer of substances and cell anchoring. *Sertoli cell processes* that penetrate into spermatid cytoplasm may be involved in morphogenesis or material transfer. *Tubulobulbar complexes* are organelle-free regions of spermatids penetrating into Sertoli cytoplasm that may anchor cells and be involved in cytoplasmic shedding. No tight junctions are found between Sertoli cells and germ cells. Adhering junctions are absent between Sertoli cells and spermatogonia and between Sertoli cells and late spermatids, thereby presumably allowing proliferation of the former and release of the latter germ cells.

The structural features of the interior of the Sertoli cell have been reviewed.[20,33,36,37] Sertoli cells possess predominantly euchromatic nuclei, prominent nucleoli and perinucleolar bodies. Sertoli cytoplasm contains aggregations of smooth endoplasmic reticulum, especially around lipid droplets and in the regions of the plasma membrane adjacent to the developing spermatid acrosomes. Secondary lysosomes, indicative of phagocytosis of degenerating sperm and residual bodies, abundant Golgi and mitochondria are also present. Unusual complexes of actin filaments sandwiched between plasma membrane and endoplasmic reticulum make up the "ectoplasmic specializations".[50] Microfilaments are abundant near the tubulobulbar processes of spermatogenic cells.[51] Microtubules, intermediate filaments and dyneins have also been described in Sertoli cells and are suggested to be involved in translocation of spermatids through the seminiferous epithelium.[50,52]

SUMMARY

The testis contains somatic and germ line cells. The developmental program of spermatogenesis begins with primordial germ cells that populate the gonad. These give rise to mitotically proliferating spermatogonia of various types including stem cells which eventually enter meiosis as spermatocytes. One round of replication followed by two cell divisions produces haploid spermatids which undergo extreme cytodifferentiation to form spermatozoa which collect in the lumen. All germ cells are closely associated with somatic Sertoli cells during their development in the seminiferous epithelium. The outer wall of the spermatogenic tubules is made of a basement lamina and a layer of peritubular myoid cells. The interstitial region between the tubules contains several somatic cell types, in particular the Leydig cells, sources of testosterone.

REFERENCES

1. Russell, L.D., Ettlin, R.A., Sinha Hikim, A.P. and Clegg, E.D. Histological and Histopathological Evaluation of the Testis. Clearwater, FL: Cache River Press, 1990.
2. Pelliniemi, L.J., Frojdman, K. and Paranko, J. Cell biology of testicular development. In: deKretser,D., ed. Molecular Biology of the Male Reproductive System. NY: Acad.Press, 1993:21-66.
3. Roberston, D.M., Risbridger, G.P., Hedger, M. and McLachlan, R.I. Growth factors in the control of testicular function. In: deKretser, D. ed. Molecular Biology of the Male Reproductive System. NY: Acad.Press, 1993:411-438.
4. Huhtaniemi, H. Hormonal control mechanisms of Leydig cells. In: deKretser, D. ed. Molecular Biology of the Male Reproductive System. NY: Acad. Press, 1993:383-410.
5. Hardy, M.P., Gelber, S.J., Zhou, Z., Penning, T.M., Ricigliano, J.W., Ganjam, V.K., Nonneman, D. and Ewing, L.L. Hormonal control of Leydig cell differentiation. Ann. N.Y. Acad. Sci. 1991; 637:152-163.
6. Vergouwen, R.P.F.A., Huiskamp, R., Bas, R.J., Roepers-Gajadien, H.L., Davids, J.A.G. and de Rooij, D.G. Postnatal development of testicular cell populations in mice. J. Reprod. Fertil. 1993; 99:479-485.
7. Pelliniemi, L.J. and Niemi, M. Fine structure of the human foetal testis. I. The interstitial tissue. Z. Zellforsch. 1969; 99:507-522.
8. Tapanainen, J., Kuopio, T., Pelliniemi, L.J. and Huhtaniemi, I. Rat testicular endogenous steroids and number of Leydig cells between the fetal period and sexual maturity. Biol. Reprod. 1984; 31:1027-1035.
9. Hahnel, A.C. and Eddy, E.M. Cell surface markers of mouse primordial germ cells defined by two monoclonal antibodies. Gamete Res. 1986; 15:25-34.
10. McCarrey, J.R., Hsu, K.C., Eddy, E.M., Klevecz, R.R. and Bolen, J.L. Isolation of viable mouse primordial germ cells by antibody-directed flow sorting. J. Exp. Zool. 1987; 242:107-111.
11. Dym, M. and Clermont, Y. Role of spermatogonia in the repair of the seminiferous epithelium following X-irradiation of the rat testis. Am. J. Anat. 1970; 128:265-282.
12. Parvinen, M., Vihko, K.K. and Toppari, J. Cell interactions during the seminiferous epithelial cycle. Inter. Rev. Cytol. 1986; 104:115-151.
13. Clermont, Y. Kinetics of spermatogenesis in mammals: Seminiferous epithelium cycle and spermatogonial renewal. Physiol. Rev. 1972; 52:198-236.
14. van Alphen, M.M.A, van de Kant, H.J.G. and de Rooij, D.G. Protection from radiation-induced damage of spermatogenesis in the rhesus monkey *Macaca mulatta* by follicle stimulating hormone. Cancer Res. 1989; 49:533-536.
15. Clermont, Y. and Rambourg, A. Evolution of the endoplasmic reticulum during rat spermiogenesis. Amer. J. Anat. 1978; 151:191-212.
16. Longo, F.J. and Anderson, E. Gametogenesis. In: Lash J. and Whittaker,

J.R. eds. Concepts of Development. Stamford, Conn.: Sinauer, 1974:3-47.
17. Franzen, A. Spermatogenesis. In: Giese, A.C., Pearse, J.S. and Pearse, V.B., eds. Reproduction of Marine Invertebrates Vol IX. Palo Alto: Blackwell/ Boxwood, 1987:1-47.
18. Leblond, C.P. and Clermont, Y. Definition of the stages of the cycle of the seminiferous epithelium in the rat. Annals N.Y. Acad. Sci. 1952; 55:548-571.
19. Clermont, Y. The cycle of the seminiferous epithelium in man. Am. J. Anat. 1963; 112:35-46.
20. Jegou, B. The Sertoli-germ cell communication network in mammals. Int. Rev. Cytol. 1993; 147:25-96.
21. Roosen-Runge, E.C. The process of spermatogenesis in mammals. Biol. Rev. 1962; 37:343-377.
22. de Rooij, D.G. Proliferation and differentiation of undifferentiated spermatogonia in the mammalian testis. In: Potten, C.S. ed. Stem Cells, Their Identification and Characterization. Edinburgh: Churchill Livingstone, 1983:87-117.
23. Tegelenbosch, R.A.J. and de Rooij, D.G. A quantitative study of spermatogonial multiplication and stem cell renewal in C3H/101F1 hybrid mouse. Mutation Res. 1993; 290:193-200.
24. Allen, D.J., Harmon, B.V. and Kerr, J.F.R. Cell death in spermatogenesis. In: Potten, C.S. ed. Perspectives on Mammalian Cell Death. London: Oxford, 1987:229-258.
25. Heller, C.G. and Clermont, Y. Kinetics of the germinal epithelium in man. Rec. Prog. Horm. Res. 1964; 20:545-575.
26. de Rooij, D.G., van Alphen, M.M.A., and van de Kant, H.J.G. Duration of the cycle of the seminiferous epithelium and its stages in the rhesus monkey *(Macaca mulatta)*. Biol. Reprod. 1986; 35:587-591.
27. Oakberg, J. A description of spermiogenesis in the mouse and its use in analysis of the cycle of the seminiferous epithelium and germ cell renewal. Am. J. Anat. 1956; 99:391-414.
28. Weinbauer, G.F. and Nieschlag, E. Hormonal control of spermiogenesis. In: deKretser, D. ed. Molecular Biology of the Male Reproductive System. NY: Acad.Press, 1993:99-142.
29. Hackstein, J.H.P. Spermatogenesis in *Drosophila.* A genetic approach to cellular and subcellular differentiation. Eur. J. Cell Biol. 1991; 56:151-169.
30. Holland, N. and Giese, A. An autoradiographic investigation of the gonads of the purple sea urchin *Strongylocentrotus purpuratus.* Biol. Bull. 1965; 128:241-258.
31. Kalt, M.R. Morphology and kinetics of spermatogenesis in *Xenopus laevis.* J. Exp. Zool. 1976; 195:393-408.
32. Wood, W.B., Hecht, R., Carr, S., Wolf, N., Hirsh, D. and Vanderslice, R. Parental effects and phenotypic characterization of mutations that affect early development in *Caenorhabditis elegans.* Dev. Biol. 1980; 74:446-469.
33. Tindall, D.J., Rowley, D.R., Murthy, L., Lipshultz, L.I. and Chang, C.H. Structure and biochemistry of the Sertoli cell. Int. Rev. Cytol. 1985; 94:127-149.
34. Stefanini, M., Conti, M., Geremia, R. and Ziparo, E. Regulatory mechanisms of mammalian spermatogenesis. In: Metz C. and Monroy, A. eds.

Biology of Reproduction. N.Y.: Acad.Press, 1985:59-102.

35. Cheng, C.Y., Zwain, I., Grima, J. and Bardin, C.W. Structure and functions of Sertoli cell proteins. In: Baccetti, B. ed. Comparative Spermatology 20 Years After. NY: Raven, 1991:617-624.
36. Jegou, B. and Sharpe, R.M. Paracrine mechanisms in testicular control. In: deKretser, D. ed. Molecular Biology of the Male Reproductive System. NY: Acad. Press, 1993:271-310.
37. Russell, L.D. and Griswold, M.D. ed. The Sertoli Cell. Clearwater, FL: Cache River Press, 1993.
38. Sertoli, E. Dell'esistenza di particolari cellule ramifacate nei canalicoli seminiferi del testicolo umano. Margagni 1865; 7:30-31.
39. Clermont, Y. Four decades of research on the biology of the male reproductive system: A few landmarks. Ann. NY Acad. Sci. 1991; 637:17-25.
40. Magre, S. and Jost, A. Sertoli cells and testicular differentiation in the rat fetus. J. Elect. Mircros. Tech. 1991; 19:172-188.
41. Fawcett, D.W. Ultrastructure and function of the Sertoli cell. In: Greep R.O. and Astwood, E.B. eds. Handbook of Physiology Sect. 7 Vol. 5. Baltimore, MD: Williams and Wilkins, 1975:21.
42. Dym, M. and Fawcett, D.W. The blood-testis barrier in the rat and the physiological compartmentation of the seminiferous epithelium. Biol. Reprod. 1970; 3:308-326.
43. Setchell, B.P. The Mammalian Testis. NY: Cornell, 1978:.
44. Gunsalus, G.L. and Bardin, C.W. Sertoli-germ cell interactions as determinants of bidirectional secretion of androgen-binding protein. Ann. N.Y. Acad. Sci. 1991; 637:327-339.
45. Vitale, R., Fawcett, D.W. and Dym, M. The normal development of the blood testis barrier and the effects of clomiphene and estrogen treatment. Anat.Rec. 1973; 176:333-344.
46. Kerr, J.B., Savage, G.N., Millar, M. and Sharpe, R.M. Response of the seminiferous epithelium of the rat testis to withdrawal of androgen: evidence for direct effect upon intercellular spaces associated with Sertoli cell junctional complexes. Cell Tiss. Res. 1993; 274:153-161.
47. McGinley, D.M., Posalaky, Z., Porvaznik, M. and Russell, L. Gap junctions between Sertoli and germ cells of rat seminiferous tubules. Tissue Cell 1979; 11:741-754.
48. Gravis, C.J. Interrelationships between Sertoli cells and germ cells in the Syrian hamster *Mesocricetus-auratus*. Z. Mikrosk. Anat. Forsch. 1979; 93:321-342.
49. Hew, K.-W., Heath, G.L., Jiwa, A.H. and Welsh, M.J. Cadmium *in vivo* causes disruption of tight junction-associated microfilaments in rat Sertoli cells. Biol.Reprod. 1993; 49:840-849.
50. Vogl, A.W., Pfeiffer, D.C. and Redenbach, D.M. Sertoli cell cytoskeleton: Influence on spermatogenic cells. In:Baccetti, B. ed. Comparative Spermatology 20 Years After. NY: Raven, 1991: 709-715.
51. Vogl, A.W. Distribution and function of organized concentrations of actin filaments in mammalian spermatogenic cells and Sertoli cells. Int. Rev. Cytol. 1989; 119:1-56.
52. Hall, E.S., Eveleth, J., Jiang, C., Redenbach, D.N. and Boekelheide, K. Distribution of the microtubule-dependent motors cytoplasmic dynein and kinesin in rat testis. Biol. Reprod. 1992; 46:817-828.

CHAPTER 2

Molecules of the Male Germ Line

INTRODUCTION

In this chapter are discussed many molecules found in the male germ line which have received recent attention. These include structural proteins, enzymes, regulatory proteins and some of unknown function. Some of these are found exclusively in the male germ line. Some have been investigated additionally or exclusively at the gene or mRNA level.

Several patterns of gene expression in the male germ line have been distinguished.[1-4] During spermatogenesis, many peptides are synthesized which are unique to the testis. Some of these are testis-specific isozymes of molecules expressed in other cell types. The transcription of many of these is believed to be regulated by testis-specific regulatory factors. Their genes should have specific cis-regulatory elements absent from their somatic cognates (see chapter 3). Some mRNAs are produced by alternative splicing. In a survey of cell-type specific gene products in mouse, twice as many spermatid-specific products were detected by in vitro translation of RNA from spermatids as from spermatocytes.[5] In addition to testis-specific gene products, other male germ line proteins are expressed at testis-specific levels.

Many genes transcribed in the testis are expressed after meiosis in the haploid spermatids. Some, perhaps all of these transcripts, may be available to sister spermatids through cytoplasmic bridges. Many male germ cell gene products are made during the late stages of spermiogenesis after RNA synthesis has ceased and thus must be encoded in stored and translationally regulated mRNAs. Testicular mRNAs not associated with polysomes have been estimated at about two-thirds of the total mRNA.[6] Various forms of post-translational regulation have also been demonstrated for germ line proteins such as by cleavage processing and secondary modifications. Thus multiple levels for regulation of gene action are available in the male germ line. These will be discussed in chapter 3.

The appearance of germ line-specific isozymes or variants may have many causes.[3,4] Use of alternate energy sources for intermediary metabolism, a variant level of activity needed for germ line function, optimization for the lower temperature of the testis, inactivation of certain regions containing the non-expressed variants of the genome during meiosis, or the evolution of genes which escape constraints on gene expression associated with spermatogenesis such as the progression of chromatin condensation. Testis-specific gene variants may have been selected because they contain cis-regulatory elements that respond to testis-specific trans-acting factors absent from the somatic variants.

I have somewhat arbitrarily divided the male germ line molecules into classes for heuristic purposes in order to facilitate access for the reader. The classes are not always mutually exclusive.

STRUCTURAL MOLECULES AND ENZYMES

ACROSOME

The acrosome is a sperm secretory granule which upon exocytosis releases or exposes proteins and enzymes with roles in gamete recognition and penetration of egg vestments. The acrosome reaction has recently been reviewed.[7] Underlying the acrosome is usually nonmuscle actin, appearing in the space between the acrosome and the nucleus.[8,9] Data so far indicate that acrosomal proteins are not made coordinately. Some seem to be under transcriptional control, others under translational control.

Proacrosin is a serine protease zymogen form of acrosin. Its gene is transcribed in pachytene spermatocytes as well as spermatids.[10-12] The glycoprotein is also present in these cell types.[13,14]

A *cathepsin-S* type protease[15] and a *cathepsin L* [16] have been reported in guinea pig acrosomes. *Cathepsin-D* type proteases have been reported in mouse[17] and rabbit.[18]

Acrogranin is a 67 kDa acrosomal acidic glycoprotein synthesized in pachytene spermatocytes and round spermatids.[19] It is incorporated into the developing acrosome of spermatids.

Intra-acrosomal protein *SP-10* is of interest as a possible candidate for a primary contraceptive vaccine. It appears on the sperm surface as a consequence of the acrosomal reaction.[20] It has been cloned from primates including humans.[21-23] Two alternatively spliced cDNAs encode 285- and 251-amino acid proteins in baboon and macaque[21] and 246- and 265-amino acid proteins in humans.[22] SP-10 is homologous to mouse protein MSA-63 which may anchor actin to the acrosomal membrane.[24] The protein has an 18-amino acid leader sequence which may direct it into endoplasmic reticulum membranes and an internal hydrophobic region. The mRNA is found mostly in round spermatids of Stages I-III and declines thereafter. The protein first appears at Stage I in round spermatids and persists throughout spermiogenesis.[25,26]

A testis-specific isoform of *protein disulfide isomerase* is found in the acrosome of developing rat spermatids but is absent from earlier germ cell stages and somatic cells of the testis.[27] Somatic PDI is usually found in endoplasmic reticulum and believed to aid in the correct folding of proteins. Its role in the acrosome is unknown. It is also found in the nucleus of spermatids during maturation and has been suggested to play a role in protamine disulfide cross-linking.

Testis membrane cofactor protein (*MCP, CD46*) is a regulatory protein of the complement system, involved in deterring bacterial infections, and is a cofactor for cleavage of some complement components by Factor I protease. It is present on acrosomal membranes and has been suggested to protect acrosome reacted sperm from complement attack in the female genital tract.[28]

Inhibition of the acrosome reaction by incubation of human sperm with pertussis toxin suggests a role for G proteins in signal transduction.[29] An α-subunit of a guanine nucleotide binding protein (*G protein*), involved in various signal transductions at the plasma membrane level, is associated with proacrosomal granules of spermatocytes and developing acrosomes of spermatids in the mouse.[30] Differential localization with development of different Gα subunits has been suggested to be involved in acrosome biogenesis or signal transduction initiating acrosome exocytosis.

Calreticulin, a major Ca^{++} binding and sequestering protein found in the endoplasmic reticulum of most tissues, is also found in the acrosome of round spermatids and spermatozoa of the rat.[31] It has 416 amino acids with a 17 residue N-terminal signal sequence, and a KDEL (lysine-aspartic acid-glutamic acid-leucine) endoplasmic reticulum signal sequence. Its mRNA is abundant in pre- and post-meiotic cells. It has two Ca^{++} binding sites, one low affinity/high capacity and the other high affinity/low capacity.

The acrosome reaction is Ca^{++}-triggered in the sea urchin, elevating sperm cAMP levels.[7] Adenyl cyclase is probably regulated by calmodulin.[32] A major sea urchin sperm acrosomal protein, which binds sperm to eggs species-specifically, is *bindin*. Bindin has been sequenced from two species of sea urchin.[33,34] It is synthesized as a precursor of 481 amino acids which is cleaved to generate a carboxyl terminal 238-amino acid protein. Bindin is well conserved especially in a central 42-amino acid domain. The more variable flanking regions presumably are responsible for species-specific binding. A putative transmembrane segment in *Arbacia*, missing from *Strongylocentrotus*, may confer membrane fusion properties upon *Arbacia* bindin. Bindin expression is completely testis-specific.[33] Bindin mRNA accumulates in haploid cells, as does the protein, which is packaged in the developing acrosome.[35]

Chromatin

Nuclear chromatin undergoes profound structural changes during spermatogenesis. For recent reviews of chromatin structural and accompanying compositional changes during spermatogenesis, see refs. 36 and 37.

The DNA of most sperm nuclei is highly condensed, sometimes approaching[37] or exceeding[38] that of close-packed nucleosome containing chromatin fibers. In some cases, the chromosomes are coiled[39] or even ordered in some way[40,41] and specific DNA sequences may occupy non-random positions. For example, specific sequences such as rDNA or centromeric satellite DNA are localized to the equatorial region of bull[42] and the posterior dorsal side of mouse sperm nuclei.[43]

Rat sperm chromatin is arranged as stacked lamellar sheets oriented parallel to the long axis of the nucleus.[44] Mouse DNA is six times as concentrated as in mitotic chromosomes.[38,45] The DNA may be anchored at a special structure called the sperm nuclear annulus lying at the implantation fossa.[45] The means by which the chromatin is reorganized and compacted during spermatogenesis is largely unknown, but its exceedingly high density is believed to be facilitated by substituting protamines for histones as the major DNA binding chromosomal proteins.[38]

Changes in chromatin accessibility and DNA backbone continuity have been detected during spermatogenesis. Using in situ nick translation, DNase 1 hypersensitive sites were found in early stages of spermatogenesis, but these disappeared during spermiogenesis in a graded fashion from anterior to posterior reflecting the progression of chromatin condensation.[46] At Steps 12-13, as transition proteins replace histones, the chromatin becomes very sensitive to nick translation even without exogenous DNase 1, and then from Steps 16-19 refractile again. It has been suggested that endogenous nicks occurring in Step 12-13 spermatids may aid in chromatin folding characteristic of this period.[46] For a review of this subject, see ref. 36.

As sperm-specific non-histone basic proteins replace typical histones in the chromatin of spermatids, transcriptional activity is silenced and nucleosomal structure is lost.[38,47] Thus haploid transcription must occur in the earlier stages of spermiogenesis.

A form of gene imprinting may take place in chromatin during spermatogenesis to indicate which genes should be immediately expressed following fertilization. Chromatin structures associated with active genes are disrupted by replacement of histones. In the chicken, constitutively expressed genes are apparently marked by hypomethylation whereas genes expressed in a tissue-specific fashion are hypomethylated later.[48] Some of these hypomethylated sites, escaping the wave of methylation occurring between spermatogonial and spermatocyte stages, correspond to sites which eventually become DNase 1 hypersensitive indicative of an altered chromatin structure associated with active genes. The role of methylation in imprinting has recently been reviewed.[49]

To what extent sperm nuclei carry signals to be used after fertilization is not known. Certainly imprinting of genes by methylation is one possibility. Nuclei are usually devoid of enzymes operating in somatic nuclei such as replication and transcription enzymes, kinases and phosphatases. Rat sperm nuclei contain small nuclear RNAs such as U1 and U2 involved in nuclear processing.[50] It is unclear whether these are left over from previously transcriptionally active stages or are to be used after fertilization.

Nuclear Proteins

During spermatogenesis in most organisms, a special set of basic nuclear proteins either replaces or supplements the chromosomal histones, a subject which has been extensively reviewed.[37,51-53] The basic proteins consist of a wide variety of molecules in different organisms and thus stand in remarkable contrast to the evolutionary conservation of histones in somatic tissues.

The classic examples of sperm-specific nuclear proteins are the protamines of bony fish discovered by Miescher over a century ago.[54-56] These molecules are small highly basic proteins (3-4 kDa, 60-70% arginine) containing several arginine clusters. In mammals, "keratinous" or "mammalian" protamines (about 7 kDa, 50-60% arginine) replace histones in mature sperm.[57-61] These resemble fish protamines in their core regions, but have additional flanking sequences, contain cysteine and are linked by disulfide bonds in the mature sperm nucleus (Fig. 2.1).[62,63] During mammalian spermiogenesis, histones of round spermatids are first replaced by transition proteins (TPs) in elongating or elongated spermatids which are then eventually replaced by protamines.

Protamine-like molecules in vertebrates have been extensively reviewed[64-66] as have male germ line-specific molecules in invertebrates.[37,52,53,66] Recent reports have characterized nuclear proteins in mature sperm of mussel,[67-69] perch,[70] several teleosts,[71,72] and various mollusks.[73] Discussions emphasizing possible evolutionary relationships of sperm-specific basic nuclear proteins may be found in refs. 55 and 74-76.

Patterns of histone replacement by germ line-specific nuclear proteins during spermatogenesis vary considerably. For example, in the sea urchin, testis-specific forms of histones (Sp H1 and Sp H2B) are present throughout spermatogenesis but in a highly phosphorylated form.[77-79] They become dephosphorylated in late spermatids to yield the forms found in mature sperm. Variations in secondary modifications during spermatogenesis are common.[64,80] These usually result in alterations of basicity and therefore interaction with DNA.[81] After fertilization, secondary modifications of these proteins precede reformation of somatic nucleohistone structure.[37,82,83]

Sperm-specific nuclear proteins in the frogs *Bufo* and *Xenopus* increase greatly in nuclei at the last steps of spermiogenesis when the chromatin begins to condense.[84] *Xenopus* sperm contain six sperm-specific proteins (SP 1-6) in addition to histones. They are synthesized at differentially regulated rates and appear at the elongating spermatid stages.[85] In synchronous cultures of round spermatids, *Xenopus* sperm-specific basic proteins were synthesized[85] whereas primary spermatocytes synthesized histones exclusively. SP4 mRNA is present in pachytene spermatocytes and thus stored prior to translation.[86] *Bufo* sperm contain only protamines, two closely related forms 39-amino acids long differing by only one amino acid,[87] which are similar to fish protamines. *Bufo* protamine-2 mRNA first appears in round spermatids.[86]

Below are discussed patterns of substitution of basic nuclear proteins during mammalian spermiogenesis.[57,60]

Testis-Specific Histones

Rodents have testis-specific variants of histones H1, H2A, H2B and H3. The rat *H1t* gene is expressed only in testis, exclusively in pachytene spermatocytes.[88] H1t is replaced completely by transition proteins at the midspermatid stages.[89] Mouse H1t has been cloned and is similar to rat and human.[90] Its mRNA is not polyadenylated. A testis-specific form of H2B histone *(TH2B)* of rat is expressed in pachytene spermatocytes but unlike typical somatic histones, in a replication-independent way. Testis-specific *TH2A* also first appears in primary spermatocytes.[91] The *H4t* gene codes for a protein identical to somatic H4 but is expressed in pachytene spermatocytes.[92] For a review of testis-specific histones, see ref. 57.

A spermatid-specific histone H2B, *SSH2B,* with an unusual hydrophobic C-terminal region, is transcribed and translated exclusively in round spermatids of mouse.[93,94] Despite its C-terminus, it is not preferentially associated with the nuclear envelope as first suggested.

Transition Proteins

Transition proteins appear in mammalian spermatids replacing histones. They are themselves replaced by protamines. *TP 1* is a 54-amino acid protein relatively well-conserved between species. *TP 2* is about twice as large and a zinc binding protein. TP 3 is now known to be a precursor to protamine-2 and TP 4 is a minor protein.[95] TPs replace histones including the highly acetylated H4 histones which are localized almost exclusively in Step 11 spermatids.[96] The replacement occurs progressively from anterior to caudal as condensation proceeds.

Amino Acid Sequence of Salmon Protamine

Ala-
Arg-Arg-Arg-Arg-Ser-Ser-Ser-Arg-Pro-Val-
Arg-Arg-Arg-Arg-Arg-Pro-Arg-Val-Ser-
Arg-Arg-Arg-Arg-Arg-Arg-Ala-Gly-
Arg-Arg-Arg-Arg-COOH

Amino Acid Sequence of Rat Protamine-1

Ala-Arg-Tyr-Arg-Arg-Cys-Cys-Arg-Ser-Lys-Ser-Arg-Ser-Arg-Cys-
Arg-Arg-Arg-Arg-Arg-Arg-Cys-
Arg-Arg-Arg-Arg-Arg-Arg-Cys-Cys-
Arg-Arg-Arg-Arg-Arg-Arg-Cys-Cys-
Arg-Arg-Arg-Arg-
Ser-Tyr-Thr-Phe-Arg-Cys-Lys-Arg-Tyr-COOH

Fig. 2.1. Comparison of the amino acid sequences of a typical teleost protamine[62] and mammalian protamine.[63] Sequences are arranged to emphasize polyarginine tracts.

TP 1 mRNA is transcribed from Step 2 until Steps 7-9 spermatids and for some time is stored in the cytoplasm not associated with any obvious cytoplasmic compartment.[97] Maximal mRNA levels are found in mouse cytoplasm at Step 9, declining by Steps 13-15.

Rat TP 2 mRNA appears by in situ hybridization in Step 7 spermatids in a perinuclear structure, probably the chromatoid body. It increases in the cytoplasm during spermiogenesis to a maximum at Steps 10-11, declining and becoming undetectable by Step 14.[98] Mouse TP 2 mRNA is first detected in Step 7 round spermatids and degraded before Step 14.[99] TP 1 and TP 2 appear in the nucleus by electron microscopic immunohistochemistry at the same time as chromatin condensation begins, but after most nuclear shaping is completed.[95] Thus they are unlikely to have a role in the latter process.

Protamines

The classic studies of protamine dynamics during spermatogenesis were performed in fish, are very well documented and will not be reviewed further here.[55,64,100-102] For a review of mammalian protamines, see ref. 60.

Mouse sperm nuclei contain two protamines, protamine-1 and protamine-2 (as do human, horse, hamster and monkey). Protamine-1 is made in its mature form; protamine-2 as a precursor. A minor form of boar protamine-1 has an additional 6-amino acid N-terminal extension apparently derived from initiation of translation at a non-AUG codon.[103] Protamine-1 is the first protamine appearing in human sperm nuclei at Step 5 increasing through Step 8 spermatids.[104] The rat makes only a protamine-1 although it has the gene for protamine-2.

Transcripts of rat protamine-1 were localized by electron microscopic in situ hybridization in the nucleus.[97] They initially appear in Steps 1-2 of spermiogenesis and decline by Steps 7-9, accumulate maximally in the cytoplasm and persist until Steps 16-19. They are not localized in any region of the cytoplasm including the chromatoid body which had been suggested to be a site of sequestration.[97]

In mouse, several precursors of the mature protamine-2 have been described. Six precursors, three of which were previously described as transition proteins, were present in the chromatin of elongating spermatids and were related as processing intermediates.[105,106] The full length precursor of 106 amino acids is cleaved in the N-terminal region at positions 4, 10, 15, 19, 25, 31 and 44 producing a 62-amino acid mature protamine-2.[107] In rat, protamine-2 mRNA is transcribed and a protein precursor is made in late spermatids.[108] Previously identified as transition protein TP 3, the protein undergoes several cleavages, but the processing is faulty and mature protamine-2 protein does not appear in the spermatozoa.[108]

In humans, there are three protamines differing in their N-terminal sequences and four related intermediate basic proteins. The intermediate basic protein-2 (HPI-2), present in low amounts in mature sperm, is structurally related to human protamine-2 (HP-2).[109] A correlation between infertility, lower HP-2 content and lesser protamine affinity for DNA has been reported.[110] Phosphorylated forms of human protamine-1 and protamine-2 persist in mature sperm. Protamine-1 is phosphorylated on Ser-10 or Ser-10 and Ser-8; protamine-2 is phosphorylated on Ser-14.[111]

The evolutionary relationships between protamine-2 precursor genes were examined by analysis of genes from various primates.[76] The precursors showed conservation of overall size and organization, but rapid mutation rates. The variability at the proteolytic sites suggests either that there are multiple processing enzymes of different specificities or that the folded structure of the molecule directs the processing sites independently of sequence specificity.[76] The genes for protamine-1, protamine-2 and TP 2 in humans are linked in a 13 kb region in the 5' $\rightarrow$ 3' direction on chromosome 16.[112] It was suggested that this cluster arose by gene duplication of the protamine-1 gene.

Lamins and Nuclear Matrix

The nuclear lamina is a proteinaceous sheath underlying the nuclear envelope which may be involved in nuclear shaping, chromatin attachment and nuclear envelope assembly. The presence or absence of a sperm nuclear lamina has been a source of controversy.

The related lamins A/C appear to be disassembled at the beginning and reassembled at the end of mitosis. Lamin B remains associated with membranes. A novel lamin B_3 exists in mice.[113] It is a 53 kDa isoform restricted to spermatogenic cells and is generated by differential splicing from lamin B_2. The gene upon transfection into somatic cells caused the spherical nuclei to change to a hook shape reminiscent of the mouse sperm nucleus. A nuclear lamina has also been reported in rat pachytene spermatocytes.[114] A rat B-type lamin of 60 kDa redistributes from the nuclear envelope to the synaptonemal complex during meiotic prophase.[115] An antibody to lamins A/C recognized a novel 52 kDa putative lamin in meiotic stages of rat.[116] Antibodies to lamin A/C did not bind to mouse male germ cells at any stage and antibodies to B bound to all stages.[117]

Xenopus lamins L_I and L_{II} are B type lamins present in spermatogonia.[118] The spermatocytes were reported to lack lamins. A male germ line-specific lamin, lamin L_{IV}, appears in *Xenopus* late in spermatogenesis and has been suggested to play a role in alteration of nuclear architecture during spermiogenesis.[118]

A nuclear matrix is also present in some sperm cells.[119] It is associated with a proliferating cell matrix antigen during rat spermatogenesis.[120]

Synaptonemal Complex Proteins

The synaptonemal complex is specific to meiosis and believed to be critical for proper chromosome pairing and crossing-over. It appears as a ladder-like structure with a central and two lateral elements and can extend at pachytene stages to the nuclear envelope at each end.[121] Slight variations in structure of the central region have been noted between insects and mammals.[122] Several proteins associated with the complex have been partially characterized. Topoisomerase II is located in the lateral elements.[123] Meiosis-specific proteins of 33 kDa and 30 kDa localize to the lateral elements;[124-126] others of 190 kDa and 130 kDa are located in the lateral elements and in the pairing region respectively.[115,127] One of 48 kDa is found in the central region of rat synaptonemal complexes.[128] Another rat component of apparently 65 kDa has been cloned.[129] The gene encodes a 50 kDa protein with potential glycosylation sites. The protein is located in the central pairing region. The gene is not transcribed testis-specifically. A protein of the synaptonemal complex in yeast, *ZIP-1,* is required for synapsis.[130] It is a putative 875-amino acid protein sharing homology with α-helical coiled coil proteins which interact hydrophobically as dimers. It has been suggested to act as a molecular zipper.

Other Nuclear Proteins

NASP (nuclear autoantigenic sperm protein), a sperm-specific nuclear protein of rabbit, has homology with *Xenopus* histone binding proteins N1/N2.[131] It is encoded by a tissue-specific, 2.5 kb mRNA and is a 74-kDa protein of 680 amino acids. The mRNA and protein appear in primary spermatocytes.[132] The protein moves from the nuclear area to the cytoplasm to the post-acrosomal region during spermatogenesis.

Laminin, a major extracellular matrix component, has been reported within nuclei of *Drosophila* male germ line cells.[133] Its transcripts are detected in spermatogonia and spermatocytes.

Other proteins, including the 60 kDa *calicin* and a set of uncharacterized related peptides, lying just outside the nucleus constitute part of the perinuclear theca.[134] This region just under the acrosomal and plasma membranes is fibrous and resistant to detergent solubilization. The proteins first appear in round spermatids. The theca probably connects to a nuclear matrix.[135]

Midpiece/Flagellum

During differentiation of the sperm, mitochondria become concentrated in the midpiece area, sometimes fusing. Mitochondria are often arranged in a helical pattern. The sperm tail axoneme extends from the midpiece region. The sperm flagellum consists of an axoneme similar to cilia which provides motility. The axoneme contains microtubular doublets of tubulin usually extending from a basal body close to the nucleus or embedded in a special region called the implantation fossa. Mammalian sperm often have an elaborate structure extending around the axoneme containing nine outer dense fibers. In the region of the annulus in the midpiece, only seven fibers are found and they are encircled by a fibrous sheath.

The mitochondrial enzyme *cytochrome c* is present as two isozymes in mammals: c_S and c_T. C_S is highly conserved and ubiquitously expressed in somatic cells as well as in Type A spermatogonia and spermatids.[136] Cytochrome c_S is especially prevalent in round spermatids and the protein has also been localized to interstitial cells, Sertoli cells, and early germ cells. Cytochrome c_T begins to replace c_S in primary spermatocytes and later stages and is the only isozyme in mature sperm.[137] They differ in 14 of the 104 amino acids. Both forms of the enzyme can be found within the same mitochondria during transition. In the rat, a 1.7 kb testicular form of c_S mRNA, larger than its somatic counterparts, results from an alternative upstream promoter.[138] In situ hybridization shows the 1.7 kb mRNA of cytochrome c_S is abundant in late meiotic prophase and in Steps 1-9 spermatids and then disappears. Cytochrome c_T mRNA first appears in zygotene spermatocytes, but is highest in round spermatids, and not present in Steps 10-19 suggesting that the protein is long-lived or that a pool of apoprotein exists in elongating spermatids.[139]

The mRNA of the second subunit of *cytochrome c oxidase*, the terminal enzyme in the mitochondrial electron transport chain, is expressed at high levels in rat pachytene spermatocytes, maximally at Stages VII-VIII.[140] It is found in much lower amounts in Sertoli cells. Expression patterns have been proposed to reflect requirements for altered energy demands during meiosis.[140]

The *E1α subunit of pyruvate dehydrogenase*, an enzyme involved in aerobic energy metabolism, is encoded by a 2.0 kb transcript which is expressed testis-specifically.[141] The mRNA is abundant in pachytene spermatocytes and earlier stages. A 1.7 kb transcript is found in round spermatids. The two messages do not have poly(A) tails of different lengths and presumably result from two different polyadenylation signals. The message is always found on polysomes and is not post-transcriptionally regulated. A somatic-type transcript is found in Leydig cells, Sertoli cells and spermatogonia, but not later germ cells.[142]

Mitochondrial capsule selenoprotein, MCS, is a cysteine-proline rich selenium containing protein located in a specialization of the outer mitochondrial membrane which may function in organizing mitochondria into a helical path in the midpiece of mouse sperm. MCS mRNA appears first in Step 3 round spermatids and remains at high levels until Step 16 at the end of spermiogenesis. Expression is restricted to haploid cells.[99]

A novel 205 kDa serine/threonine protein kinase, *MAST205*, a microtubule-associated protein of the manchette was cloned from mouse testis.[143] The microtubular manchette forms around the base of the round spermatid nucleus and may play a role in nuclear shaping. The mRNA is expressed maximally in spermatids.

A sperm-specific form of α-tubulin appears in late spermatids of mouse during manchette and axoneme formation.[144] It is only expressed in haploid cells.

Drosophila has a testis-specific form of *tubulin β2.*[145] Tubulin β1 is expressed in somatic testicular cells and early germ cells and is completely replaced in germ cells by β2 during meiosis. Transformation of the germ line with β1 and β3 subtypes under control of the β2 regulatory regions shows that all three forms can co-assemble.[145] The carboxyl end defines isotype and is not required for assembly. The β2 carboxyl end is required for proper organization of the axoneme although microtubules assemble in its absence.[146]

Proteins of the rat outer dense fibers share epitopes with Sertoli cell secreted protein *S70.*[147] S70 is an acidic glycoprotein. The dense fibers contain keratin-like fibrous proteins. It is not clear whether spermatids make molecules sharing common epitopes with the Sertoli products or whether the Sertoli products are precursors to the sperm antigens.

A protein similar to tektin, an axonemal protein found in the wall of flagellar microtubules, has been found to bind to the *kl-3* loop on the Y lampbrush chromosome of primary spermatocytes in *Drosophila.*[148] The *Drosophila* Y chromosome has not yet been shown to encode functional polypeptides.[149] It was proposed that the Y loops function in protein binding that may be involved in correct assembly of the axoneme.[148]

CYTOPLASM

A variety of other cytoplasmic proteins have been investigated in the male germ line. They are arranged according to function below.

Metabolic and Other Enzymes

Several enzymes of intermediary metabolism are found in unique forms in the testis. A testis-specific form of *lactate dehydrogenase,* LDH-C_4, is a tetrameric enzyme in mouse. It appears in preleptotene spermatocytes, is present in round spermatids and decreases in elongated spermatids.[150-152] Five isozymes are found in somatic cells assembled from the gene products of LDH A and B in different combinations of the two polypeptides. The testicular form is more sensitive to pyruvate inhibition and less to lactate inhibition, perhaps adapted to the high concentrations of lactate in oviduct fluid, and this variant is better at oxidizing lactate.[153] Regulation of expression differs between mouse and rat. Mice have much higher levels of mRNA than rat due to a somewhat higher rate of transcription, but no difference in cytoplasmic stability is found, suggesting post-transcriptional differences, probably in processing or nuclear stability.[154]

Phosphoglycerate kinase (PGK) is an enzyme of the glycolytic pathway. The somatic and a testis-specific form are expressed during spermatogenesis. The proteins and mRNAs for Pgk-1 (somatic) or for Pgk-2 forms are expressed in parallel suggesting that they are primarily controlled at the transcriptional level.[6,155,156] Pgk-2 mRNA is first seen in preleptotene spermatocytes increasing through the round spermatid stage, whereas Pgk-1 declines in pachytene spermatocytes.[155]

Hexokinase, the first enzyme in the glycolytic pathway has three unique germ cell-specific mRNAs coding for proteins nearly identical to the somatic form.[157] In their 5' regions, each shows both a unique sequence and another in common to the three. Two are developmentally regulated during spermatogenesis.

Superoxide dismutase, SOD, an enzyme which scavenges reactive oxygen species, was localized to human spermatogonia and somatic cells of the epididymis and prostate.[158] SOD mRNA is present in the testis as 0.77 and 0.94 kb transcripts, the latter developmentally regulated, peaking at 60 days in rats.[159] In situ hybridization localized the mRNA in seminiferous tubules of prepubertal rats, but could not distinguish germ cell from Sertoli expression.[159]

Hypoxanthine phosphoribosyltransferase (hprt) is an X-linked gene expressed throughout spermatogenesis despite X-chromosome inactivation, although the level of its activity declines. In the absence of an autosomal counterpart, this suggests that either the gene escapes inactivation or the gene products are stabilized after inactivation.[160]

Lecithin-retinol acyltransferase increases 10-fold in specific activity between days 20 and 35 in rat testis due to post-meiotic germ cells. It may be used to make retinyl esters from retinol received via Sertoli cells.[161]

A temperature-labile form of *cholesteryl ester hydrolase (CEH)* has been purified from rat testis.[162] It is probably an isozyme unique to testis which may regulate intracellular cholesterol pools. βCEH protein levels increase at the beginning of spermatogenesis. Four more thermostable forms are found in Sertoli and Leydig cells.

A testis-specific form of *angiotensin-converting enzyme (ACE)*, an enzyme which can regulate blood pressure, is expressed abundantly in germ cells. Though smaller than its somatic cognate, it is enzymatically active. Its testicular function is unknown. Alteration of two residues by site-directed mutagenesis (Lys-154 and Tyr-236) previously shown by chemical alteration to be important in enzyme function, changed rabbit ACE catalytic abilities substantially.[163] Alteration of human ACE Tyr-200 (corresponding to Tyr-236 of rabbit) also altered catalytic properties.[164]

Phosphatases and Kinases

Protein kinase A (PKA) is a cyclic AMP dependent protein kinase implicated in many regulatory circuits involving cAMP. Four isoforms of the regulatory subunit R are differentially expressed during spermatogenesis.[165,166] Two are smaller forms produced by alternative polyadenylation.[166] RIβ expression is maximal at Stages VIII-XI in late pachytene spermatocytes and round spermatids; RIIα is maximal at Stages VIIc,d-VIII in round and elongating spermatids; RIIβ in Stages VII-VIII in round spermatids; and RIα is essentially constant in pachytene spermatocyte and round spermatids suggesting differential functions for the isozymes.

Adenyl cyclase converts ATP to cAMP and is found in a testis-specific form in the mouse male germ line.[167,168] Since adenylate cyclase appears to be constitutively synthesized in male germ cells, it has been suggested that levels of cAMP are dependent on catabolic control through phosphodiesterases.[169] *cAMP phosphodiesterase (cAMP-PDE)* converts cAMP to AMP. Two forms are expressed in rat testis.[170] Rat PDE1 is encoded in a 4.0 kb and in a 2.8 kb transcript which are abundant in pachytene spermatocytes, decrease in round spermatids and are absent from condensing spermatids and residual bodies. PDE2 is encoded in 4.0, 3.5, 2.8, 2.4 kb transcripts which are abundant in round spermatids and less so in condensing spermatids/residual bodies. An enzyme activity peak A present in all germ cells may be PDE2; peak B present in pachytene spermatocytes and round spermatids may be PDE1.

Three of eight known isotypes of *protein kinase C* are expressed in mouse testis. The transcripts for the PKC-δ and -θ forms are about 500 bp shorter than their somatic transcript counterparts; the -ζ form is unchanged. The 2.2 kb transcript encoding the Ca^{++}-independent protein kinase C-θ is expressed predominantly in mouse testis.[171]

Mak protein kinase (*Mak*) is related to cdc2 M-phase kinase in its kinase domain. Its mRNA is restricted to post-meiotic testicular cells, especially late pachytene spermatocytes and declines in spermatids. It exists in the cytoplasm as a 66 and a 60 kDa protein associated with a 210 kDa protein which it phosphorylates.[172] The kinase mRNAs are expressed at high levels at Stages IV-VI in round spermatids in retinol synchronized rat testes, but are suppressed just after retinol replacement.[172] Retinol down regulates the 2.4 and 3.4 kb Mak transcripts during spermatogonial proliferation following retinol replacement. Mak transcripts respond differently to retinol in early germ cells and in spermatocytes or round spermatids suggesting two different pathways of retinol signaling.[173]

Nek1 is a protein kinase expressed at high levels in spermatocytes and round spermatids.[174] It is a 774-amino acid protein which is 42% identical in its N-terminal kinase domain with NIMA, a kinase from *Aspergillus nidulans* which controls the initiation of mitosis. It is both a serine/threonine and a tyrosine kinase.

A unique catalytic subunit of *calmodulin-dependent phosphatase (CaM-PrP)* cloned from a cDNA library has a predicted molecular weight of 59 kDa and is expressed in testis as a 2.8 kb transcript in later stages of spermiogenesis.[175] It may exist as a flagellar form associated with regulation of sperm motility[176] and may become associated with a *calcineurin B-like protein.*[177] Calcineurin is phosphatase 2B, a Ca^{++}/calmodulin stimulated protein phosphatase containing two subunits: α, a calmodulin-binding catalytic subunit, and β, a Ca^{++} binding regulatory subunit. A regulatory subunit of calcineurin, CaNβ1, exists in rat testis as a testis-specific isoform expressed from the beginning of meiosis throughout spermiogenesis.[178]

Serine-proline protein phosphatases exhibit a complex pattern of expression in testis. The β variant of the B regulatory subunit of *phosphatase PP-2A* from rat testis has been partially cloned.[179] It is expressed in rat brain and testis, particularly in elongated spermatids. The catalytic subunit of PP-2A is calcineurin. Other phosphatases have been cloned and characterized. The α-isotype is expressed in all stages of spermatogenesis. PP-1γ2 mRNA and protein are specific to spermatocytes and early spermatids and located in nuclei.[180] The transcript is produced by alternative splicing with PP-1γ1. PP-2AβS, a catalytic subunit, is specific to testis and is the same protein as PP-2AβL. Theirin RNAs differ in the length of their poly(A) tails.[181]

Protein Degradative Enzymes

The chicken *UbII polyubiquitin* gene is expressed preferentially during spermiogenesis.[182] Ubiquitin is a 76-amino acid globular protein which is involved in protein degradation. In testis, use of alternative initiation sites within the 5' intron of the embryonic form makes ubiquitin synthesis insensitive to heat shock. It was postulated that the testis-specific form is necessary for germ cell differentiation at the elevated temperature of birds. It is expressed in premeiotic cells, meiotic cells and early spermatids.

Furin and *PC4* are processing endopeptidases resembling yeast *Kex-2.* Furin is found in germ and somatic cells, PC4 in only round spermatids of the mouse. It has been suggested that PC4 may process proenkephalin transcripts from round spermatids.[183,184]

Cytoskeletal Proteins

Most *actin* in male germ cells is in the intercellular bridges and the subacrosomal space.[185] In spermatozoa, it is usually concentrated in the posterior head and neck region.[185] Actin and calmodulin co-localize in the subacrosomal region of spermatids.[8,186] Immunochemically, *ankyrin* localizes around the neck and *spectrin* around the fibrous sheath of the flagellum.[187]

Cytoplasmic actin expression is high at the onset of mitotic activity and muscle actin expression parallels muscle sheet expansion in starfish. A putative acrosomal actin is expressed at high levels in spermiogenesis.[188]

The major sperm protein *MSP* of the nematode *C. elegans* is a unique cytoskeletal motility element for its non-flagellated sperm. It has been recently crystallized.[189] The gene *spe-6* is required to assemble MSP for proper motility.[190]

Ca^{++}-Binding Proteins

Calmodulin, a Ca^{++}-dependent regulatory protein, is localized by electron microscopy to the nucleus and cytoplasm of pachytene spermatocytes and round spermatids in the mouse where it appears to saturate calmodulin binding proteins.[191] It is present at very low levels in spermatogonia and other testicular cells. It is eventually localized to the post-acrosomal region in epididymal sperm. Some calmodulin binding proteins accumulate in nuclei towards the end of spermatogenesis, as revealed by a calmodulin overlay assay, but calmodulin itself does not increase during this period.

Calbindin-D is a 28 kDa vitamin D-dependent Ca^{++} binding protein found immunochemically in spermatogonia and spermatocytes of the chick.[192] It has been suggested to play a role in mitosis based upon the developmental time of its appearance.

Heat Shock Proteins

Heat shock proteins are synthesized in response to external stress and mediate protein-protein interactions. *Hsp86* mRNA and protein levels are high in meiotic prophase and in spermatids. Their parallel expression suggests that hsp86 is not translationally regulated.[193] *Hsp84* was expressed in testicular somatic cells also not apparently translationally controlled. These forms coprecipitate with hsp70. An *hsp70* protein of 630 amino acids called *hsc70t* has been cloned from mouse. The protein is localized to late spermatids, residual bodies and mature sperm, and the mRNA appears in early spermatids.[194] It is transcriptionally activated in round spermatids.[195] A heat shock-like protein *hst-70* from rat homologous to hsp70.2 from mouse is expressed in pachytene spermatocytes.[196-199]

Plasma Membrane and Surface

Many differentiation antigens of the mammalian germ line have been detected. These are grouped as early-class, late-class, and transient antigens by developmental time of appearance.[200-202] These antigens often occupy discrete domains on the spermatozoan cell surface, and may undergo extensive rearrangement during spermatogenesis.[203]

Four surface domains on mature epididymal guinea pig sperm exhibiting antigen localization are the anterior head, posterior head, anterior tail and posterior tail regions.[203] In the testis, four distribution patterns of antigens have been discriminated: whole cell, whole head, anterior tail and posterior tail.

Immunochemical detection of antigen can be due to insertion of the molecule into the plasma membrane, or unmasking or modification of an existing epitope. Loss of antigen can be due to removal, masking or destruction. Some antigens have been shown to be inserted or removed from the surface.[203] Guinea pig *PH-20* surface antigen in round spermatids appears first in the Golgi apparatus and then in the acrosomal membrane, but later goes directly to the plasma membrane.[204] Both forms are uniformly distributed initially, only later restricted, most likely through mechanisms involving partial proteolysis[205] and barriers to diffusion within the membrane.[206] Antigen *PH-30*, involved in gamete fusion, is an integral membrane glycoprotein which exists as a dimer. Each subunit is cleaved from a larger precursor.[207] The pre-α subunit is present during spermatogenesis and processed by the time of spermatozoon formation. Pre-β is processed in the epididymis.

A transiently expressed differentiation antigen, a 95 kDa glycoprotein, is present on mouse zygotene and early pachytene spermatocytes.[201] Intraperitoneal injection of a monoclonal antibody against this antigen caused disturbances of spermatogenesis.

In sea urchin, several surface antigens show different spatial patterns of localization during spermatogenesis.[208] The timing of localization is also variable. One antigen, a 210 kDa glycoprotein believed to act as either a Ca^{++} channel or channel regulator and restricted to the germ line, appears in cytoplasmic vesicles at all stages of spermatogenesis and appears on the surface of mature sperm cells.[209]

REGULATORY MOLECULES

PROTO-ONCOGENES

The cellular homologues of RNA tumor virus oncogenes are involved in signal transduction and cascades which control proliferation and differentiation. They code for growth factors and their receptors, signal transducers including receptor and non-receptor triggered tyrosine kinases, cytoplasmic serine/threonine kinases, GTP binding proteins and nuclear factors including transcriptional regulators.[210] See ref. 211 for a recent review and list of oncogenes.

Many proto-oncogenes are expressed in the male germ line and most await functional studies. In spermatogenesis, oncogenes could potentially regulate commitment of stem cells to differentiate, progress of meiosis, initiation of stage-specific transcription, signal transductions mediating hormone or growth factor responses, or spermiogenesis-specific differentiation. Although stage-specific expression of certain proto-oncogenes such as maximal activity in spermatocytes or exclusive activity in spermatids has led to suggestions for particular roles, little is yet certain. Some like *c-myc*, *c-jun* and *c-fos* are abundant in B-Type spermatogonia, some such as *junD*, *N-ras* or *c-abl* are detected as haploid-specific transcripts, others such as *H-ras* are expressed in late pachytene spermatocytes and early spermatids, and others like *K-ras* and *c-mos* predominate in pachytene spermatocytes with little activity in spermatids.

c-Abl codes for a non-receptor serine-threonine kinase of 150 kDa. All tissues of mouse express the gene with two transcripts of 5.3 and 6.5 kb. Testis is the most abundant site of expression in the adult with a 4.7 kb transcript in addition to the two somatic types.[212,213] The 4.7 kb form encodes a protein of similar size to the somatic version with deleted sequences from the 3' untranslated region which may help stabilize it in spermatids.[213a,214,215] It is most abundant in late round and elongating spermatids.[216] Unfortunately no obvious defects were found in transgenic mice homozygous for mutant *c-abl*.[217,218]

A testis-specific transcript in mouse encodes a truncated *fer* protein, *ferT*, a plasma membrane associated kinase. The 2.4 kb transcript codes for the smallest known tyrosine kinase, 453-amino acids long, which is nearly identical to the carboxyl halves of rat and human *fer* proteins.[219] The truncated version is suggested to result from alternative splicing. It is expressed at high levels in early spermatids preceding the expression of several other oncogenes such as *int-1, c-mos, and c-abl*.[219]

Pim-1 encodes a serine/threonine kinase that may be involved in terminal differentiation. It is expressed as a unique 2.4 kb transcript in spermatids.[220] This transcript results from an alternative polyadenylation signal which confers increased stability.[221]

c-Mos is a cytoplasmic serine-threonine kinase expressed as a unique sized testicular transcript of 1.7 kb found in low levels in round spermatids.[216,222] The highest levels of protein are found in pachytene spermatocytes.[223] Its promoter shares a sequence in common with several other testis-specific genes.[224] Since *c-mos* has been identified as the cytostatic factor in *Xenopus* eggs, presumably operating by blocking cyclin degradation, it may play a similar role in controlling proliferation in the male germ line.

The *raf* oncogenes are serine-threonine cytoplasmic kinases whose roles in spermatogenesis remain to be elucidated. All *raf* products have conserved central portions with cysteine-rich zinc finger regions, serine/threonine-rich domains which may serve for autophosphorylation, and kinase domains. *C-raf-1* is expressed in all somatic and germ cells of the testis as a 3.3 kb-transcript, highest in early pachytene spermatocytes and lowest in spermatogonia.[225] MAP kinase is a substrate for *C-raf-1* in vivo.[226] *A-raf* is expressed mostly in somatic cells as a 4.3 kb transcript, especially in Leydig cells, suggesting possible paracrine regulation. A 2.3-kb form is probably expressed in the germ line.[225] *A-raf* can activate AP-1 transcription factor. *B-raf* is a 75 kDa protein expressed testis-specifically as 4.0 and 2.6 kb trancripts in spermatids.[225] Because, unlike *C-raf-1* and *A-raf*, it has a consensus sequence for cdc2 kinase phosphorylation just before its zinc finger region, it may be regulated by cyclin/p34cdc2.

The *ras* oncogenes are GTPases involved in transducing surface signals to MAP kinase and through a kinase cascade ultimately to transcription factors regulating gene expression. They are 21 kDa proteins which bind and hydrolyze GTP. *H-ras* (1.3 and 1.1 kb transcripts) is expressed in meiotic stages and in spermatids, *K-ras* (2.2 kb) predominantly in pachytene spermatocytes but not spermatids, and *N-ras* in spermatogonia (5.0 kb) and spermatocytes (1.4 kb).[220,227] Several other GTP-binding proteins are expressed are in testis such as *rab1, rab2, rab4, ralA* and *rho12*[165]. In addition, the Gi subunit of G proteins has been localized in pachytene spermatocytes and round spermatids immunohistochemically.[228]

β-*Chimaerin* is a GTPase-activating protein (GAP) which regulates GTPases. GTPases cycle between active forms with GTP bound and inactive forms with GDP bound. GAPs break down the active form. β-Chimaerin is found as a 30 kDa protein in rat testis, expressed at late stages of spermatogenesis during assembly of the acrosome. It has a phorbol ester receptor and a *rho/rac* GAP domain indicating it prefers the *rac* subfamily of *ras* type GTPases. The *rac* subfamily is involved in actin polymerization.[229] A *Drosophila rac*-GAP is expressed in primary spermatocytes.[230] In *Drosophila,* the *rotund* locus *rn* codes for a protein like human *rac* GTPase-activating protein and is expressed in primary spermatocytes.[230]

Wnt-1 encodes a secretory glycoprotein believed to function in paracrine and autocrine regulation. The product is 36-44 kDa. *Wnt-1* is expressed in the adult as a testis-specific 2.6 kb transcript in post-meiotic cells.[231] Transgenic mice with antisense constructs of *wnt-1* to avoid early lethality of "knockout" constructs and turned on with the meiotic PGK-2 promoter have expression suppressed by 98%, but show little effects on fertility or testicular histology suggesting that the gene is not essential for spermatogenesis. This raises the possibility that some post-meiotically expressed genes are aberrant transcripts resulting during chromatin reorganization during spermiogenesis.[232] Other *wnt* genes of unknown function are *wnt5b* and *wnt6.*[233]

Several proto-oncogenes act in nuclei to regulate transcription. Early response genes such as *c-myc, c-fos,* and *c-jun* can respond to mitogenic stimuli by activating transcription of other genes.

c-fos mRNA was reported to be maximal in Type B spermatogonia,[234] but this may be an artifact of cell dissociation. The 2.2 kb transcript produces a 62 kDa protein. Interaction with *c-jun* produces the AP-1 transcription factor.

c-Jun is also maximally expressed in Type B spermatogonia as a 2.7 kb transcript.[234] In addition, family members *junB* and *junD* are expressed in mouse testis,[235] especially in Type B spermatogonia suggesting that they may also contribute to regulation of mitotic division. Levels of *c-jun* and *junB* proteins decline from preleptotene spermatocytes to background at the pachytene stage,[236,237] but *jun D* remains high in meiotic and post-meiotic cells.

c-Myb from *Xenopus* is a member of a family of *mybs*, nuclear proteins which act as transcriptional regulators. They have a conserved DNA-binding domain repeat of about 50 amino acids containing regularly spaced tryptophan residues. The protein is found mostly in spermatogonial nuclei and down-regulated during meiosis.[238]

c-Myc is a 58 kDa nuclear phosphoprotein implicated in cell proliferation which may act as a transcription or replication activator. Its transcript is maximally expressed in Type B spermatogonia.[234] *c-Myc* mRNA was reported in human sperm cells in the post-acrosomal midpiece and tail regions.[239] Since *c-myc* mRNA has a short half life it may function in mature sperm. *c-Myc* protein has also been reported in the acrosomal region of mature sperm.[240] A *myc* from starfish, *pAv-myc,* resembling mammalian *myc* with domains for transcriptional activation, casein kinase II phosphorylation, nuclear targeting, basic DNA binding and oligomerization, varies during the seasonal reproductive cycle and is activated during spermatogonial mitoses.[241]

Ets-2 encodes a 3.5 kb transcript in prepubertal mice. The 56 kDa nuclear protein, like its counterpart in *Xenopus* eggs, may be a transcription factor involved in meiosis.[242] *Ets-1* is expressed in immature testis as a 5.3 kb transcript.[242] A related gene *elk-1* is expressed as a 1.7 kb transcript.[243]

c-ErbA is a transcriptional regulator and thyroid hormone T_3 receptor. A testis-specific form *c-erbA-T* encodes a 55 kDa protein in humans.[244] A form generated by alternative splicing which does not bind T_3 can bind to the T_3 response element, and so may function as a repressor.[245,246]

The *c-kit* oncogene is discussed in chapter 6.

Tumor Suppressor Genes

The tumor suppressor genes inhibit cell proliferation and may have a role in regulation of stem cell cycling. The retinoblastoma *rb* tumor suppressor gene, is expressed as a 4.7 kb transcript in all adult tissues, a unique transcript in post-meiotic cells between pachytene spermatocyte and early spermatid stages.[247] The unphosphorylated form of *rb* blocks exit from G1 of the cell cycle. In testis, the gene product may be modulated by binding of differentiation-regulated transcription factor 1.[248]

The *p53* tumor suppressor gene acts at the G1/S border to inhibit proliferation. The protein has a C-terminal DNA-binding domain and an acidic N-terminal transcription domain. It is involved in apoptosis and cell differentiation and may halt DNA replication. It has been found in a normal and an alternatively spliced shorter form in testis which differ in their C-terminal regions.[249] In mouse testis, its mRNA is found exclusively in pachytene spermatocytes and the protein is abundant in mid-pachytene spermatocytes thus suggesting a role in meiotic control.[250] A hybrid transgene, a p53 promoter-CAT construct, was expressed mostly in the testis and in primary spermatocytes.[251] Transgenic mice with a high multiplicity of p53 promoter-CAT constructs exhibited low levels of p53 protein presumably due to competition for transcription factors with endogenous genes.[252] These mice exhibited giant-cell degenerative syndrome ascribed to an inability of primary spermatocytes to complete meiotic division. Instead they undergo further replications to give giant multinucleated cells.[252] It has been speculated that the function of p53 in spermatogenesis is to permit sufficient time for DNA repair and reshuffling to occur during meiosis.[250]

Homeobox Genes

Motifs often associated with transcription activation proteins are helix-turn-helix, helix-loop-helix, zinc finger motifs, HMG boxes and homeoboxes. Regulatory aspects of some of these are discussed in chapter 3.

Homeobox genes contain a conserved-183 bp region originally identified in homeotic *Drosophila* genes which regulate development. *Hox* gene products bind sequence-specifically to DNA and often function as transcriptional regulators.[253,254] They are involved in complex networks of gene regulation by interacting with *hox* and other genes to control levels of expression.

In adult mouse testis, *hox 1.4* is uniquely expressed at high levels in spermatocytes and round and elongating spermatids.[255] The gene encodes a long untranslated 3' region, and its two transcripts of 1.35 and 1.45 kb are smaller than their embryonic counterparts. Expression is abolished in *at/at* mutants which lack germ cells. The gene is expressed appropriately in transgenic mice. Deletion analysis shows at least one potential *cis* regulatory element between 2-4 kb 5' to the embryonic transcription start site.[256]

Several other homeobox genes are expressed in testis including *hox 1.1, hox 1.2,*[257,258] *hox 1.3,*[259] *hox 1.6,*[260] *hox 2.6* and *hox 4.2.*[261,262] *Hox* genes are differentially regulated. Whereas *hox* 1.4 mRNA is at high levels only in testis, *hox 2.6* is expressed at low levels in germ cells and somatic cells of adult testis, and *hox 4.2* is restricted to somatic interstitial cells.[263] *Hox 1.3* and *1.5* are expressed in germ cells at lower levels. *Hox 1.3* produces a unique transcript of 1.6 kb and *hox 1.5* a single unique transcript of 2.7 kb.[263]

Octamer binding proteins (Oct family) contain a POU domain conferring high affinity DNA binding and are homeobox-type transcription factors which recognize the consensus sequence ATGCAAAT and may regulate tissue-specific transcription. Many Oct proteins are expressed in testis of mouse: *Oct-1, Oct-2A* and *-2B, mini Oct-2, Oct-4, Oct-6, Oct-8, Oct-9, Oct-11* and *Brn-3.*[264-272]All are expressed in meiotic cells and *Oct-1* and *Oct-4* are also expressed in primordial germ cells. *Mini Oct-2* is a truncated version of *Oct-2* that may negatively regulate genes controlled positively by other Oct proteins.[269]

Zinc Finger Proteins

The mouse *Zfy-1* and *Zfy-2* and human *ZFY* genes encode zinc-finger proteins originally thought to be involved in primary sex determination. *Zfy-1* has 13 zinc fingers and an acidic region;[273] *Zfy-2* lacks part of this region and has a disrupted third zinc finger.[274,275] Adult expression is limited to the male germ line.[275,276] *Zfy-1* is initially expressed in primordial germ cells and peaks at the beginning of meiosis. *Zfy-2* is also abundant during meiosis.

Several testicular zinc finger proteins have been isolated from adult mouse libraries using a probe for zinc finger proteins. *Zfp-29* in adult mice is expressed exclusively in the testis, in spermatocytes and spermatids, especially round spermatids.[276] *Zfp-35* is expressed testis-specifically, especially in pachytene spermatocytes and spermatids.[277,278]

Zfp-37 is expressed only in testis of adults as 2.3 and 2.6 kb major transcripts which are most abundant in spermatids.[279] It encodes a 553-amino acid protein with a basic 216-amino acid, serine/threonine-rich N-terminal domain and a C-terminal region containing 12 zinc fingers in tandem.

Zfp-38 is a strong transcriptional activator of 555 amino acids containing seven zinc fingers in the 197-amino acid C-terminal region, and a transacting domain in the non-finger region.[280] Its mRNA is expressed from pachytene spermatocytes to round spermatids.

Rex-1, another zinc finger protein, is regulated by retinoic acid and expressed in spermatocytes.[281] The zinc finger proteins *MOK2,*[282] *Krox-8,*[283] *Krox-20,*[284] *mKr3, mKr4, mkr5,*[285] and *TSGA*[286] are all expressed in testis but the cell type is uncertain. *Krox-20* binds to the 5' flanking region of homeobox gene *hox 1.4.*[264]

HMG Box Proteins

HMG box proteins contain regions of homology to high mobility group chromosomal proteins and are found in several sequence-specific DNA binding proteins and transcriptional regulators. Two motifs of the HMG box were found in a gene, isolated from a screen of a mouse testicular DNA library, containing a 12 bp element from the PGK-2 promoter.[287] The 23 kDa protein, *tsHMG,* is expressed mostly in Steps 9 and 10 spermatid nuclei at the time of transcription cessation and chromatin condensation,. Another HMG box protein is encoded by *SRY,* a gene involved in mammalian sex-determination[288] (see chapter 6).

Cell Cycle and Replication Genes

A cyclin B type gene, *cycB1,* expressed in several adult tissues is found in testis as a 1.7 kb transcript expressed in elongating spermatids, pachytene spermatocytes and especially in round spermatids. Its function in non-cycling cells may be to regulate phosphorylation of histones or other nuclear proteins or microtubules.[289] *Cyclin B2,* also a 1.7 kb transcript, is expressed maximally in pachytene spermatocytes[290] and to a lesser extent in early round spermatids. Curiously, neither cyclin is expressed in cycling spermatogonia. In human sperm, cyclin A and B and cdc2 kinase were detected immunologically in the acrosome region.[291] Antibodies to the three proteins increased the acrosome reaction and it was suggested that the proteins were involved in capacitation or acrosome reaction.

Ornithine decarboxylase (ODC) is an enzyme needed for cell proliferation which controls a rate-limiting step in polyamine synthesis. ODC mRNA levels are high in pachytene spermatocytes and round spermatids,[292,293] but the protein decreases during the time when its mRNA increases.[294] Most ODC mRNA is non-polysomal suggesting translational control.[292,293] In addition, a low molecular weight cytosolic factor may be responsible for post-translational control.[294] Two transgenic mouse lines overexpressing human ODC contained high levels of putrescine.[295] One showed enhanced replication of Type A_4, intermediate and Type B spermatogonia; the other only Type B spermatogonia with inhibition of meiotic DNA synthesis. These results suggest a role for ODC in positive and negative regulation of DNA synthesis in mitotic and meiotic cycles of spermatogenesis.

DNA polymerase-β is involved in DNA repair. Transcripts of 1.3 and 1.4 kb are found in the germ line, the 1.3 kb mostly in leptotene and zygotene spermatocytes and the 1.4 kb mostly in spermatids. Both transcripts are predominantly non-polysomal in spermatocytes, but switch to polysomes in round spermatids correlating with their protein expression.[296]

Poly(ADP)ribose polymerase is also involved in repair synthesis. Transcripts of 3.7 and 3.8 kb are present in Type A and Type B spermatogonia, in high amounts in spermatocytes where they are on polysomes and in low amounts in spermatids where they are not on polysomes. Most of the protein is localized to the nuclei of pachytene spermatocytes.[296]

Prohibitin, a negative regulator of growth, is expressed in leptotene spermatocytes with only low levels in preleptotene and zygotene spermatocytes.[297] It is expressed maximally at Stages VII-XI of the rat. Three germ cell transcripts of 1.5, 1.2 and 0.7 kb are found, the smaller two not present in somatic testicular cells. Only a single 30 kDa protein is expressed. Its pattern of expression suggests that it may be involved in blocking mitotic and meiotic cell division.

OTHER PRODUCTS

Mouse *laminin receptor* proteins are found by immunoblotting in germ cells. Since these are not located near basement lamina the receptor may have other functions.[298] The mRNAs are 1.2 kb in testis and all somatic tissues, and 1.3 and 0.9 kb exclusively in round spermatids. The size difference between 1.3 and 1.2 kb is due to poly(A) tail length, whereas the 0.9 kb transcript uses a different start site or is differentially processed.

Epidermal growth factor precursor is synthesized in post-meiotic germ cells of the mouse.[299] The precursor, EGFp, is found in pachytene spermatocytes and round spermatids. The mature form, EGFm, is found in Sertoli cells, pachytene spermatocytes and round spermatids.

A *glucocorticoid receptor (GR)* component, a 94 kDa protein of low steroid affinity, was found in high concentrations in epididymal mouse sperm, but lower in pachytene spermatocytes and round spermatids suggesting that it accumulates late in spermatogenesis.[300] Its germ line function is unknown. A complex of the 90 kDa heat shock protein *hsp90* and this molecule is required for high affinity ligand binding, but hsp90 is absent from late cells, suggesting that the high affinity receptor is probably not assembled into receptor complexes in these cells.[300] The 6 kb glucocorticoid receptor mRNA is found in round spermatids.[300] Other investigators have immunolocalized GR in primary spermatocytes as well as in Leydig and other interstitial cells, but not spermatozoa, and have proposed a direct effect of GR on spermatogenesis and steroidogenesis.[301]

Retinoic acid receptor mRNA is found in germ cells, increasing at Stages VII-IX.[302] It may mediate the dramatic effects of vitamin A deprivation on spermatogenesis.

A number neural peptides involved in intercellular signaling or their mRNAs are expressed in the male germ line.[303] *Cholecystokinin*[304] and *gastrin*[305] are found in the acrosome, *glutamic acid decarboxylase* (an enzyme in the γ-amino butyric acid pathway) and *choline acetyltransferase* (an enzyme involved in acetyl choline synthesis) are expressed in the germ line.[306,307] *Proenkephalin* may be stored in the sperm acrosome.[308] *Vasopressin* and *oxytocin* are also expressed in testis.[309]

Tpx-1, a testis-specific gene, encodes a product sharing homology with acidic epididymal glycoprotein (AEG) and may be involved in sperm maturation. It is transcribed at high levels in haploid cells.[310] It contains a signal peptide sequence but no transmembrane domain.

A mouse *RNA helicase P68,* an ATP-dependent RNA processing enzyme involved in splicing, ribosome assembly, and translation initiation,[311] is expressed in late pachytene spermatocytes and haploid spermatids.[312] Its amino acid sequence is 95% identical to the human helicase and it contains a signature DEAD (asp-glu-ala-asp) box. A related helicase, *PL10,*[313] is maximally expressed in meiotic cells and spermatids. Since it is homologous to translation initiation factor elF-4A, it has been proposed to regulate translation of cytoplasmic transcripts.

Rb97D is a *Drosophila* protein that binds to hnRNA and is expressed throughout development. Its probable function is in RNA processing, but it is not known if it is specific or general. Multiple transcripts are formed by alternative splicing, generating two protein isoforms. The gene is first expressed in the testis of third instar larvae in spermatocytes and spermatids but not in earlier cells. A mutant with a P-element inserted into the 5' untranslated region demonstrates that it is necessary for male fertility.[314] The mutation resulted in the complete absence of sperm in an otherwise apparently normal testis. Germ cells up to elongating spermatids appeared, followed by degeneration of later cells.

A monoclonal antibody raised against a 60-70 kDa human sperm protein, *YWK-II,* cross reacts with rat sperm. The rat gene has homology with the cytoplasmic and transmembrane domains of the A4 amyloid precursor protein of Alzheimer's disease.[315] The gene is expressed in all stages of spermatogenesis.

Megl is a gene expressed during meiosis in mice which was isolated from an adult cDNA library. An abundant transcript of 0.75 kb encodes a 10.8 kDa lysine-rich protein containing many phosphorylation sites. It is only expressed in testis, particularly in pachytene spermatocytes and abruptly declines in spermatids unlike most other genes commonly showing spermatocyte expression.[316]

The *Mea-2 male-enhanced antigen* gene was selected from a cDNA testis library using a monoclonal histocompatibility antibody to the male-specific HY protein. It is expressed in pachytene spermatocytes and round spermatids. *Mea-1,* isolated using a polyclonal serum against the HY-antigen, is similar. The Mea gene products may code for serological HY-antigens encoded by the Y chromosome.[317]

The *mouse Y box protein 1 (MSY1)* gene encodes a testis-enriched mRNA first detected in pachytene spermatocytes where it is maximally transcribed and translated.[318] The protein is found as an mRNP particle of 60-80 S, does not show sequence or structure specificity in binding and is hypothesized to regulate germ cell message storage and translation. It may have a second role in transcriptional regulation. Proteins showing homology to the *Xenopus* FRG *Y2* box RNA binding proteins have been reported in mouse male germ cells.[319] A Y-chromosome linked gene family *YRRM* with RNA-binding protein homology may encode proteins involved in RNA processing or translational control in humans[320] and other mammals. The genes are expressed testis-specifically and deletions in the RNA recognition motif have been correlated with oligospermia. The genes of this family are candidates for the azoospermia factor gene *AZF* causing male infertility.

Other Y genes include *TSPY* in humans which may have a role in sperm differentiation.[321] It has a bovine but no known mouse homologue. *Spy* in mouse is required for proliferation of spermatogonia and has no known human homologue.[322] Y353B may be involved in spermiogenesis.[323]

SUMMARY

The male germ line expresses many molecules tissue-specifically. Some of these are isotypes of somatically expressed forms. Others are unique to the testis. Some genes are transcribed in the haploid spermatids. Some proteins are only translated well after their transcription. Structural proteins unique to the spermatozoon are of interest as products of extreme differentiation, so called " luxury" molecules. Some isoforms of common "housekeeping" enzymes may be better adapted to the conditions of the testis. Male germ line cells are the only known eukaryotic cells which can dispense with the highly conserved chromosomal histones. Regulatory molecules are of interest for their abilities to control patterns of gene expression or integrate extracellular signals and cellular responses. Regulation of germ line gene expression is discussed in chapter 3.

REFERENCES

1. Wolgemuth, D.J. and Watrin, F. List of cloned mouse genes with unique expression patterns during spermatogenesis. Mamm. Genome 1991; 1:283-288.
2. Erickson, R.P. Post-meiotic gene expression. Trends Genet. 1990; 6:264-269.
3. Willison, K. and Ashworth, A. Mammalian spermatogenic gene expression. Trends Genet. 1987; 3:351-355.
4. Hecht, N.B. Gene expression during male germ cell development. In: Desjardins, C. and Ewing, L.L. eds. Cell and Molecular Biology of the Testis. Oxford: Oxford Univ. Press, 1992:464-503.
5. Fujimoto, H. and Erickson, R.P. Functional assays for mRNA detect many new messages after male meiosis in mice. Biochem. Biophys. Res. Comm. 1982; 108:1369-1375.
6. Gold, B., Fujimoto, H., Kramer, J.M., Erickson, R.P. and Hecht, N.B. Haploid accumulation and translational control of phosphoglycerate kinase-2 mRNA during mouse spermatogenesis. Devel. Biol. 1983; 98:392-399.

7. Hardy, D.M. and Garbers, D.L. Molecular basis of signalling in spermatozoa. In: deKretser, D. ed. Molecular Biology of the Male Reproductive System. NY: Acad.Press, 1993:233-270.
8. Fouquet, J.-P. and Kann, M.-L. Species-specific localization of actin in mammalian spermatozoa: Fact or fiction?. Micro. Res. Tech. 1992; 20:251-258.
9. Oko, R., Hermo, L. and Hecht, N.B. Distribution of actin isoforms within cells of the seminiferous epithelium of the rat testis: Evidence for a muscle form of actin in spermatids. Anat. Rec. 1991; 230:489-501.
10. Kashiwabara, S., Baba, T., Takada, M. Watanabe, K., Yano, Y. and Arai, Y. Primary structure of mouse proacrosin deduced from the cDNA sequence and its gene expression during spermatogenesis. J. Biochem. 1990; 108:785-791.
11. Kashiwabara, S., Arai, Y., Kodaira, K. and Baba, T. Acrosin biosynthesis in meiotic and postmeiotic cells. Biochem. Biophys. Res. Comm. 1991; 173:240-245.
12. Kremling, H., Keime, S., Wilhelm, K., Adham, I.M., Hameister, H. and Engel, W. Mouse proacrosin gene: nucleotide sequence, diploid expression and chromosomal localization. Genomics 1991; 11:828-834.
13. Anakwe, O., Sharma, S., Hardy, D.M. and Gerton, G.L. Guinea pig proacrosin is synthesized principally by round spermatids and contains O-linked as well as N-linked oligosaccharide side chains. Molec. Reprod. Devel. 1991; 29:172-179.
14. Escalier, D., Gallo, J.-M., Albert, M., Meduri, G., Bermudez, D., David, G. and Schrevel, J. Human acrosome biogenesis: Immunodetection of proacrosin in primary spermatocytes and of its partitioning pattern during meiosis. Development 1991; 113:779-788.
15. McDonald, J.K., Culbertson, J.T. and Owers, N.O. Identification and localization of a novel cathepsin S-like proteinase in guinea pig spermatozoa. Arch. Biochem. Biophys. 1993; 305:1-8.
16. McDonald, J.K., and Kadkhodayan, S. Cathepsin L - A latent proteinase in guinea pig sperm. Biochem. Biophys. Res. Comm. 1988; 151:827-835.
17. Erickson, R.P. and Martin, S.R. The relationship of mouse spermatozoal to mouse testicular cathepsins. Arch. Biochem. Biophys. 1974; 165: 114-120.
18. Srivastava, P.N. and Ninjoor, V. Isolation of rabbit testicular cathepsin D and its role in the activating of proacrosin. Biochem. Biophys. Res. Comm. 1982; 109:63-69.
19. Anakwe, O.O. and Gerton, G.L. Acrosome biogenesis begins during meiosis. Evidence from the synthesis and distribution of an acrosomal glycoprotein acrogranin during guinea-pig spermatogenesis. Biol. Reprod. 1990; 42:317-328.
20. Herr, J.C., Flickinger, C.L., Homyk, M., Klotz, K. and John, E. Biochemical and morphological characterization of the intraacrosomal antigen SP-10 from human sperm. Biol. Reprod. 1990; 42:181-193.
21. Freemerman, A.J., Wright, R.M., Flickinger, C.J. and Herr, J.C. Cloning and sequencing of baboon and cynomolgus monkey intraacrosomal protein SP-10: Homology with human SP-10 and a mouse sperm antigen MSA-63. Molec. Reprod. Devel. 1993; 34:140-148.
22. Wright, R.M., John, E., Klotz, K., Flickinger, C.J. and Herr, J.C. Cloning and sequencing of cDNAs coding for the human intraacrosomal anti-

gen SP-10. Biol. Reprod. 1990; 42:693-701.
23. Wright, R.M., Suri, A.K., Kornreich, B., Flickinger, C.J. and Herr, J.C. Cloning and characterization of the gene for the human acrosomal protein SP-10. Biol. Reprod. 1993; 49:316-325.
24. Liu, M.S., Ruedi, A. and Lee, C.Y.G. Molecular and developmental studies of a sperm antigen recognized by HS-63 monoclonal antibody. Biol. Reprod. 1992; 46:937-948.
25. Kurth, B.E., Wright, R.M., Flickinger, C.J. and Herr, J.C. Stage-specific detection of mRNA for the sperm antigen SP-10 in human testes. Anat. Rec. 1993; 236:619-625.
26. Kurth, B.E., Klotz, K., Flickinger, C.J. and Herr, J.C. Localization of sperm antigen SP-10 during the six stages of the cycle of the seminiferous epithelium in man. Biol. Reprod. 1991; 44:814-821.
27. Ohtani, H., Wakui, H., Ishino, T., Komatsuda, A. and Miura, A.B. An isoform of protein disulfide isomerase is expressed in the developing acrosome of spermatids during rat spermiogenesis and is transported into the nucleus of mature spermatids and epididymal spermatozoa. Histochem. 1993; 100:423-429.
28. Cervoni, F., Fenichel, P., Akhoundi, C., Hsi, B.-L. and Rossi, B. Characterization of a cDNA clone coding for human testis membrane cofactor protein MCP, CD46. Molec. Reprod. Devel. 1993; 34:107-113.
29. Lee, M.A., Check, J.H. and Kopf, G.S. A guanine nucleotide-binding regulatory protein in human sperm mediates acrosomal exocytosis induced by the human zona pellucida. Mol. Reprod. Devel. 1992; 31:78-86.
30. Karnik, N.S., Newman, S., Kopf, G.S. and Gerton, G.L. Developmental expression of G protein α subunits in mouse spermatogenic cells: Evidence that Gαi is associated with the developing acrosome. Devel. Biol. 1992; 152:393-402.
31. Nakamura, M., Moriya, M., Baba, T., Michikawa, Y., Yamanobe, T., Arai, K., Okinaga, S. and Kobayashi, R. An endoplasmic reticulum protein, calreticulin, is transported into the acrosome of the rat sperm. Exp. Cell Res. 1993; 205:101-110.
32. Bookbinder, L.H., Moy, G.W. and Vacquier, V.D. Identification of sea urchin sperm adenyl cyclase. J. Cell Biol. 1990; 111: 1859-1866.
33. Gao, B., Klein, L.E., Britten, R.J., and Davidson, E.H. Sequence of mRNA coding for bindin, a species-specific sea urchin sperm protein required for fertilization. Proc. Natl. Acad. Sci. USA 1986; 83:8634-8638.
34. Glabe, C.G. and Clark, D. The sequences of the *Arbacia punctulata* bindin cDNA and implications for the structural basis of species-specific sperm adhesion and fertilization. Devel. Biol. 1991; 143:282-288.
35. Nishioka, D., Ward, R., Poccia, D., Kostacos, C. and Minor, J.E. Localization of bindin expression during sea urchin spermatogenesis. Molec. Reprod. Devel. 1990; 27:181-190.
36. McPherson, S. and Longo, F.J. Chromatin structure-function alterations during mammalian spermatogenesis: DNA nicking and repair in elongating spermatids. Eur. J. Histochem. 1993; 37:109-128.
37. Poccia, D. Remodelling of nucleoproteins during gametogenesis, fertilization, and early development. Int. Rev. Cytol. 1986; 105:1-65.
38. Pogany, G.C., Corzett, M., Weston, S., and Balhorn, R. DNA and protein content of mouse sperm. Exp. Cell Res. 1981; 136:127-136.

39. Amemiya, S., Uemura, I., Sinkai, T. and Sato, H. A birefringence, electron microscopy, and histochemical survey of chromosomal structure in the sperm nuclei of an echinothurid sea urchin, *Araeosoma owstoni.* Exp. Cell Res. 1988; 179:446-453.
40. Inoue, S. and Sato, H. Arrangement of DNA in living sperm: a biophysical analysis. Science 1962; 136:1122-1124.
41. Suzuki, M. and Wakabayashi, T. Packaging of DNA in cricket sperm: a compact mode of DNA packaging. J. Mol. Biol. 1988; 204:653-661.
42. Powell, D., Cran, D.G., Jennings, C. and Jones, R. Spatial organization of repetitive DNA sequences in the bovine sperm nucleus. J. Cell Sci. 1990; 97:185-191.
43. Powell, D., Cran, D.G., Jennings, C. and Jones, R. Spatial organization of the genome in the mammalian sperm nucleus. In: Baccetti, B. ed. Comparative Spermatology. 20 Years After. New York: Raven, 1991:67-71.
44. Koehler, J.K., Wurschmidt, U. and Larsen, M.P. Nuclear and chromatin structure in rat spermatozoa. Gamete Res. 1992; 8:357-370.
45. Ward, W.S. and Coffey, D.S. DNA packaging and organization in mammalian spermatozoa: comparison with somatic cells. Biol. Reprod. 1991; 44:569-574.
46. McPherson, S.M. and Longo, F.J. Localization of DNAse I-hypersensitive regions during rat spermatogenesis: Stage-dependent patterns and unique sensitivity of elongating spermatids. Molec. Reprod. Devel. 1992; 31:268-279.
47. Biggiogera, M., Muller, S., Courtens, J.L., Fakan, S. and Romanini, M.G.M. Immunoelectron microscopical distribution of histones H2B and H3 and protamines in the course of mouse spermiogenesis. Micro. Res. Tech. 1992; 20:259-267.
48. Groudine, M. and Conkin, K.F. Chromatin structure and *de novo* methylation of sperm DNA: Implications for activation of the paternal genome. Science 1985; 228:1061-1068.
49. Eddy, E.M., Welch, J.E. and O'Brien, D.A. Gene expression during spermatogenesis. In: deKretser, D. ed. Molecular Biology of the Male Reproductive System. NY: Acad.Press, 1993:181-232.
50. Concha, I.I., Urzua, U., Yanez, A., Schroeder, R., Pessot, C. and Burzio, L.O. U1 and U2 snRNA are localized in the sperm nucleus. Exp. Cell Res. 1993; 204:378-381.
51. Bloch, D.P. A catalog of sperm histones. Genetics Supp. 1969; 61:93-111.
52. Bloch, D.P. Histones of sperm. In: King, R.C ed. Handbook of Genetics Vol. 5. New York: Plenum, 1976:139-167.
53. Kasinsky, H.E. Specificity and distribution of sperm basic proteins. In: L. Hnilica, G. Stein and J. Stein. In, Histones and Other Basic Nuclear Proteins. Boca Raton, FL: CRC, 1989:73-163.
54. Miescher, F. Die spermatozoan einiger wirbelthieren. In: Die Histochemischen and Physiologischen Arbeiten von Friedrich Miescher. Leipzig: Vogel Vlg., 1897:359-410.
55. Dixon, G.H., Aiken, J.M., Jankowski, J.M., McKenzie, D.I., Moir, R. and States, J.C. Organization and evolution of the protamine genes of salmonid fishes. In: Reeck, G., Goodwin, G.H. and Puigdomenech, P. eds. Chromosomal Proteins and Gene Expression. N.Y.: Plenum, 1993:287-314.
56. Seniuk, N.A., Tatton, W.G., Cannon, P.D., Garber, A.T. and Dixon,

G.H. First expression of protamine message in trout testis. Ann. NY Acad. Sci. 1991; 637:277-288.

57. Meistrich, M.L. Histone and basic nuclear protein transitions in mammalian spermatogensis. In: Hnilica, L.S., Stein, G.S. and Stein, J.L. eds. Histones and Other Basic Nuclear Proteins. Boca Raton, FL: CRC Press, 1989:166-182.
58. Hecht, N.B. Mammalian protamines and their expression. In: Hnilica, L.S., Stein, G.S. and Stein, J.L., eds. Histones and Other Basic Nuclear Proteins. Boca Raton, FL: CRC Press, 1989:348-374.
59. Hecht, N.B. Molecular biology of structural chromosomal proteins of the mammalian testis. In: K.W. Adolph, ed. Molecular Biology of Chromosome Function. NY: Springer Verlag, 1989:396-421.
60. Balhorn, R. Mammalian protamines: structure and molecular interactions. In: Adolph, K.W. ed. Molecular Biology of Chromosome Function. New York: Springer-Verlag, 1993:366-395.
61. Kistler, W.S. Structures of testis-specific histones, spermatid transition proteins, and their genes in mammals. In: Hnilica, L.S., Stein, G.S. and Stein, J.L., eds. Histones and Other Basic Nuclear Proteins. Boca Raton, FL: CRC Press, 1989:332-346.
62. Moir, R.D. and Dixon, G.H. Characterization of a protamine gene from the chum salmon *Oncorhynchus-keta.* J. Mol. Evol. 1988; 27:8-16.
63. Ammer, H. and Henschen, A. Rat sperm protamine isolation and sequence analysis. Biol. Chem. Hoppe-Seyler 1988; 369:1301-1306.
64. Oliva, R. and Dixon, G.H. Vertebrate protamine genes and the histone-to-protamine replacement reaction. Prog. Nucl. Acid Res. Molec. Biol. 1991; 40:25-94.
65. Oliva, R. and Dixon, G.H. Expression and processing of the rooster protamine mRNA. Ann. NY Acad. Sci. 1991; 637:289-299.
66. Chevaillier, P. Nuclear protein transitions during sperm differentiation. In: Baccetti, B. ed. Comparative Spermatology 20 Years After. NY: Raven, 1991:19-25.
67. Carlos, S., Hunt, D.F., Rocchini, C., Arnott, D.P. and Ausio, J. Post-translational cleavage of a histone H1-like protein in the sperm of *Mytilus.* J. Biol. Chem. 1993; 268:195-199.
68. Ruiz-Lara, S., Prats, E., Casas, M.T. and Cornudella, L. Molecular cloning and sequence of a cDNA for the sperm-specific protein ϕ-1 from the mussel *Mytilus edulis.* Nucl. Acids Res. 1993; 21:2774.
69. Carlos, S., Jutglar, L., Borrell, I., Hunt, D.F. and Ausio, J. Sequence and characterization of a sperm-specific histone H1-like protein of *Mytilus californianus.* J. Biol. Chem. 1993; 268:185-194.
70. Chao, H. and Davies, P.I. Amino acid sequence of the unique protamine from yellow perch. FEBS Lett. 1992; 299:166-168.
71. Saperas, N., Lloris, D. and Chiva, M. Sporadic appearance of histones, histone-like proteins and protamines in sperm chromatin of bony fish. J. Exp. Zool. 1993; 265:575-586.
72. Saperas, N., Ribes, E., Buesa, C., Garcia-Hegart, F. and Chiva, M. Differences in chromatin condensation during spermiogenesis in two species of fish with distinct protamines. J. Exp. Zool. 1993; 165:185-194.
73. Daban, M., Chiva, M., Rosenberg, E., Kasinsky, H.E. and Subirana, J.A. Protamines in prosobranchian gastropod mollusca vary with different modes

of reproduction. J. Exp. Zool. 1991; 257:265-283.

74. Kasinsky, H.E., Mann, M., Lemke, M. and Huang, S.-Y. Diversity of sperm basic chromosomal proteins in the vertebrates: A phylogenetic point of view. In: Reeck, G.R. Goodwin, G.A. and Puigdomenech, P., eds. Chromosomal Proteins and Gene Expression. NY: Plenum, 1993:333-355.
75. Kasinsky, H.E. Sperm basic protein diversity and the biology of fertilization. In: Baccetti, B. ed. Comparative Spermatology 20 Years After. NY: Raven, 1991:41-44.
76. Retief, J.D. and Dixon, G.H. Evolution of pro-protamine P2 genes in primates. Eur. J. Biochem. 1993; 214:609-615.
77. Poccia, D.L. and Green, G.R. Packaging and unpackaging the sea urchin genome. Trends Biochem. Sci. 1992; 17:223-227.
78. Poccia, D., Lieber, T. and Childs, G. Histone gene expression during sea urchin spermatogenesis: an in situ hybridization study. Molec. Reprod. Devel. 1989; 1:219-229.
79. Poccia, D.L., Simpson, M.V. and Green, G.R. Transitions in histone variants during sea urchin spermatogenesis. Devel. Biol. 1987; 121: 445-453.
80. Massey, D.B. and Watts, S.A. Modified forms of histones H3 and H4 exhibit seasonal patterns in testes of the sea star *Asterias vulgaris.* Comp. Biochem. Physiol. 1993; 104B:181-186.
81. Green, G.R., Lee, H.-J. and Poccia, D.L. Phosphorylation weakens DNA binding by peptides containing multiple "SPKK" sequences. J. Biol. Chem. 1993; 268:11247-11255.
82. Poccia, D.L. Reactivation and remodelling of the sperm nucleus following fertilization. In: Schatten, H. and Schatten, G. eds. The Molecular Biology of Reproduction. NY: Acad.Press, 1989:115-139.
83. Green, G.R. and Poccia, D. Phosphorylation of sea urchin sperm H1 and H2B histones precedes chromatin decondensation and H1 exchange during pronuclear formation. Devel.Biol. 1985; 108: 235-245.
84. Moriya, M. and Katagiri, C. Immunoelectron microscopic localization of sperm-specific nuclear basic proteins during spermatogenesis in anuran amphibians. Devel. Growth Differ. 1991; 33:19-27.
85. Abe, S.-I. and Hiyoshi, H. Synthesis of sperm-specific basic nuclear proteins SPs in cultured spermatids from *Xenopus laevis.* Exp. Cell Res. 1991; 194:90-94.
86. Mita, K., Takamune, K. and Katagiri, C. Genes for sperm-specific basic nuclear proteins in *Bufo* and *Xenopus* are expressed at different stages in spermatogenesis. Devel. Growth Diff. 1991; 33:491-498.
87. Takamune, K., Nishida, H., Takai, M. and Katagiri, C. Primary structure of toad sperm protamines and nucleotide sequence of their cDNAs. Eur. J. Biochem. 1991; 196:401-406.
88. Grimes, S.R., Wolfe, S.A., Anderson, J.V., Stein, G.S. and Stein, J.L. Structural and functional analysis of the rat testis-specific histone H1t gene. J. Cellul. Biochem. 1990; 44:1-17.
89. Grimes, S.R., Wolfe, S.A. and Koppel, D.A. Tissue-specific binding of testis nuclear proteins to a sequence element within the promoter of the testis-specific histone H1t gene. Arch. Biochem. Biophys. 1992; 296:402-409.
90. Drabent, B., Bode, C. and Doenecke, D. Structure and expression of the mouse testicular H1 histone gene H1t. Biochim. Biophys. Acta 1993;

1216:311-313.
91. Trostle-Weige, P.K., Meistrich, M.L., Brock, W.A., Nishioka, K.and Bremer, J.W. Isolation and characterization of TH2A, a germ cell-specific variant of histone 2A in rat testis. J. Biol. Chem. 1982; 257:5560-5567.
92. Grimes, S., Weisz-Carrington, P., Daum, H., Smith, J., Green, L., Wright, K., Stein, G. and Stein, J. A rat histone H4 gene closely associated with the testis-specific H1t gene. Exp. Cell Res. 1987; 173:534-545.
93. Moss, S.B. and Orth, J.M. Localization of a spermatid-specific histone H2B protein in mouse spermiogenic cells. Biol. Reprod. 1993; 48: 1047-1056.
94. Moss, S.B., Challoner, P.B. and Groudine, M. Expression of a novel histone 2b during mouse spermiogenesis. Devel. Biol. 1989; 133:83-92.
95. Alfonso, P.J. and Kistler, W.S. Immunohistochemical localization of spermatid nuclear transition protein 2 in the testis of rats and mice. Biol. Reprod. 1993; 48:522-529.
96. Meistrich, M.L., Trostle-Weige, P.K., Lin, R., Bhatnagar, Y.M. and Allis, C.D. Highly acetylated H4 is associated with histone displacement in rat spermatids. Molec. Reprod. Devel. 1992; 31:170-181.
97. Morales, C., Kwon, Y.K. and Hecht, N.B. Cytoplasmic localization during storage and translation of the mRNAs of transition protein 1 and protamine 1, two translationally regulated transcripts of the mammalian testis. J. Cell Sci. 1991; 100:119-131.
98. Saunders, P.T.K., Millar, M.R. Maguire, S.M. and Sharpe, R.M. Stage-specific expression of rat transition protein 2 mRNA and possible localization to the chromatoid body of step 7 spermatids by in situ hybridization using a nonradioactive riboprobe. Molec. Reprod. Develop. 1992; 33:385-391.
99. Shih, D.M. and Kleene, K.C. A study by in situ hybridization of the stage of appearance and disappearance of the transition protein 2 and the mitochondrial capsule selenoprotein mRNAs during spermatogenesis in the mouse. Molec. Reprod. Develop. 1992; 33:222-227.
100. Sanders, M.M. and Dixon, G.H. The biosynthesis of protamine in trout testis. J. Biol. Chem. 1972; 247:851-855.
101. Sung, M.T. and Dixon, G.H. Modification of histones during spermiogenesis in trout: A molecular mechanism for altering histone binding to DNA. Proc. Natl. Acad. Sci. USA 1970; 67:1616-1623.
102. Dixon, G.H. The basic proteins of trout testis chromatin: Aspects of their synthesis, post-synthetic modifications and binding to DNA. In: Karolinska Symposium on Research Methods in Reproductive Endocrinology Vol.5 Stockholm: Karolinska Intitutet, 1972: 130-154.
103. Pirhonene, A., Jaaskelainen, S., Linnala-Kankkunen, A. and Maenpaa, P.H. An aminoterminally extended P1-protamine variant in the boar. Biochem. Biophys. Res. Comm. 1993; 196:831-836.
104. Lescoat, D., Blanchard, Y., Lavault, M.T., Quernee, D. and LeLannou, D. Ultrastuctural and immunocytochemical study of P1 protamine localization in human testis. Andrologia 1993; 25:93-99.
105. Elsevier, S.M., Noiran, J. and Carre-Eusebe, D. Processing of precursor of protamine P2 in mouse. Eur. J. Biochem. 1991; 196:167-175.
106. Carre-Eusebe, D., Lederer, F., Le, K.H.D. and Elsevier, S.M. Processing of the precursor of the protamine P2 in mouse. Peptide mapping and

N-terminal sequence analysis of intermediates. Biochem. J. 1991; 277:39-45.
107. Chauviere, M., Martinage, A., DeBarle, M., Sautiere, P. and Chevaillier, P. Molecular characterization of six intermediate proteins in the processing of mouse protamine P2 precursor. Eur. J. Biochem. 1992; 204: 759-765.
108. Bunick, D., Balhorn, R., Stanker, L.H. and Hecht, N.B. Expression of rat protamine 2 gene is suppressed at the level of transcription and translation. Exp. Cell Res. 1990; 188:147-152.
109. Alimi, E., Martinage, A., Arkhis, A., Belaiche, D., Sautiere, P. and Chevaillier, P. Amino acid sequence of the human intermediate basic protein 2 HPI2 from sperm nuclei. Structural relationship with protamine P2. Eur. J. Biochem. 1993; 214:445-450.
110. Belokopytova, I.A., Kostyleva, E.I., Tomilin, A.N. and Vorob'ev, V.I. Human male infertility may be due to a decrease of the protamine P2 content in sperm chromatin. Molec. Reprod. Develop. 1993; 34:53-57.
111. Chirat, F., Arkhis, A., Martinage, A., Jaquinod, M., Chevaillier, P. and Sautiere, P. Phosphorylation of human sperm protamines HP1 and HP2: identification of phosphorylation sites. Biochim. Biophys. Acta 1993; 1203:109-114.
112. Nelson, J.E. and Krawetz, S.A. Linkage of human spermatid-specific basic nuclear protein genes. Definition and evolution of the P1->P2->TP2 locus. J.Biol. Chem. 1993; 268:2932-2936.
113. Furukawa, K. and Hotta, Y. cDNA cloning of a germ cell specific lamin B3 from mouse spermatocytes and analysis of its function by ectopic expression in somatic cells. EMBO J. 1993; 12:97-106.
114. Vester, B., Smith, A., Krohne, G. and Benevente, R. Presence of a nuclear lamina in pachytene spermatocytes of the rat. J. Cell Sci. 1993; 104:557-563.
115. Heyting, C., Dietrich, A.J.J., Moens, P.B., Dettmers, R.J., Offenberg, H.H., Redeker, E.J.W. and Vink, A.C.G. Synaptonemal complex proteins. Genome 1989; 31:81-87.
116. Smith, A. and Benavente, R. Identification of a short nuclear lamin protein selectively expressed during meiotic stages of rat spermatogenesis. Differ. 1992; 52:55-60.
117. Moss, S.B., Burnham, B.L. and Bellve, A. The differential expression of lamin epitopes during mouse spermatogenesis. Molec. Reprod. Devel. 1993; 34:164-174.
118. Benavente, R. and Krohne, G. Change of karyoskeleton during spermatogenesis of *Xenopus:* Expression of lamin L_{IV}, a nuclear lamina protein specific for the male germ line. Proc. Natl. Acad. Sci. USA 1985; 82:6176-180.
119. Farrington, J.E., Patel, R. and Ward, W.S. DNA attachment to the hamster sperm nuclear matrix and nuclear annulus. Ann. N.Y. Acad. Sci. 1991; 637:164-173.
120. Hadjiolova, K.V., Martinova, Y.S., Yankulov, K.Y., Davidoc, V., Kancheva, L.S. and Hadjiolova, A.A. An immunocytochemical study of the proliferating cell nuclear matrix antigen p125/6.5 during rat spermatogenesis. J. Cell Sci. 1989; 93:173-177.
121. van Wettstein, D., Rasmussen, S.W. and Holm, P.B. The synaptonemal

complex in genetic segregation. Ann. Rev. Genet. 1984; 18:331-431.

122. Schmekel, K., Skoglund, U. and Daneholt, B. The three-dimensional structure of the central region in a synaptonemal complex - A comparison between rat and two insect species, *Drosophila melanogaster* and *Blaps cribrosa.* Chromosoma 1993; 102:682-692.
123. Moens, P.B. and Earnshaw, W.C. Anti-topoisomerase II recognizes meiotic chromosome cores. Chromosoma 1989; 98:317-322.
124. Heyting, C., Moens, P.B., van Raamsdonk, W., Dietrich, A.J.J., Vink, A.C.G. and Redeker, E.J.W. Identification of two major components of the lateral elements of synaptonemal complexes in the rat. Eur. J. Cell Biol. 1987; 43:148-154.
125. Heyting, C., Dettmers, R.J., Dietrich, A.J.J., Redeker, E.J.W. and Vink, A.C.G. Two major components of synaptonemal complexes are specific for meiotic prophase nuclei. Chromosoma 1988; 96:325-332.
126. Moens, P.B., Heyting, C., Dietrich, A.J.J., van Raamsdonk, W. and Chen, Q. Synaptonemal complex antigen localization and conservation. J. Cell Biol. 1987; 105:93-103.
127. Offenberg, H.H., Dietrich, A.J.J. and Heyting, C. Tissue distribution of two major components of synaptonemal complexes of the rat. Chromosoma 1991; 101:83-91.
128. Smith, A. and Benavente, R. Identification of a structural protein component of rat synaptonemal complexes. Exp. Cell Res. 1992; 198:291-297.
129. Chen, Q., Pearlman, R.E. and Moens, P.B. Isolation and characterization of a cDNA encoding a synaptonemal complex protein. Biochem. Cell Biol. 1992; 70:1030-1038.
130. Sym, M., Engebrecht, J. and Roeder, G.S. ZIP1 is a synaptonemal complex protein required for meiotic chromosome synapsis. Cell 1993; 72:365-378.
131. Welch, J., Zimmerman, L.J., Joseph, D.R. and O'Rand, M.G. Characterization of a sperm-specific nuclear autoantigenic protein I. Complete sequence and homology with the *Xenopus* protein, N1/N2. Biol. Reprod. 1990; 43:559-568.
132. Welch, J.E. and O'Rand, M.G. Characterization of a sperm-specific nuclear autoantigenic protein. II. Expression and localization in the testis. Biol. Reprod. 1990; 43:569-578.
133. Wang, F., Hanske, M., Miedema, K., Klein, G., Ekblom, P. and Henning, W. Laminin in the male germ cells of *Drosophila.* J. Cell Biol. 1992; 119:977-988.
134. Longo, F.J., Krohne, G. and Franke, W.W. Basic proteins of the perinuclear theca of mammalian spermatozoa and spermatids: A novel class of cytoskeletal elements. J. Cell Biol. 1987; 105: 1105-1120.
135. Bellve, A.R., Chandrika, R., Martinova, Y. and Barth, A. The perinuclear matrix as a structural element of the mouse sperm nucleus. Biol. Reprod. 1992; 47:451-465.
136. Hake, L.E., Alcivar, A.A. and Hecht, N.B. Changes in mRNA length accompany translational regulation of the somatic and testis specific cytochrome c genes during spermatogenesis in the mouse. Development 1990; 110:249-257.
137. Goldberg, E., Sberna, D., Wheat, T.E., Urbanski, G.J. and Margoliash, E. Cytochrome c: immunofluorescence localization of the testis-specific

form. Science 1977; 196:1010-1012.
138. Hake, L.E. and Hecht, N.B. Utilization of an alternative transcription initiation site of somatic cytochrome c in the mouse produces a testis-specific cytochrome c mRNA. J. Biol. Chem. 1993; 268:4788-4797.
139. Morales, C.R., Hake, L.E. and Hecht, N.B. Cellular localization of the mRNAs of the somatic and testis-specific cytochromes c during spermatogenesis in the rat. Molec. Reprod. Develop. 1993; 34:196-205.
140. Saunders, P.T.K., Millar, M.R., West, A.P. and Sharpe, R.M. Mitochondrial cytochrome c oxidase II messenger ribonucleic acid is expressed in pachytene spermatocytes at high levels and in a stage dependent manner during spermatogenesis in the rat. Biol. Reprod. 1993; 48:57-67.
141. Iannello, R.C. and Dahl, H.-H. M. Transcriptional expression of a testis-specific variant of the mouse pyruvate dehydrogenase E1α subunit. Biol. Reprod. 1992; 47:48-58.
142. Takakubo, F. and Dahl, H.-H.M. The expression pattern of pyruvate dehydrogenase E1α subunit genes during spermatogenesis in adult mouse. Exp. Cell Res. 1992; 199:39-49.
143. Walden, P.D. and Cowan, N.J. A novel 205-kilodalton testis-specific serine/threonine protein kinase associated with microtubules of the spermatid manchette. Mol. Cell. Biol. 1993; 13:7625-7635.
144. Distel, R.J., Kleene, K.C. and Hecht, N.B. Haploid expression of a mouse testis α-tubulin gene. Science 1984; 224:68-70.
145. Glatzer, K.H., Kaltschmidt, B. and Renkawitz-Pohl, R. Cell-type specific expression of β1-, β2- and β3-tubulin during spermatogenesis of *Drosophila melanogaster* and *D. hydei.* In: Baccetti, B. ed. Comparative Spermatogenesis 20 Years After. NY: Raven, 1991: 727-731.
146. Fackenthal, J.D., Turner, F.R. and Raff, E.C. Tissue-specific microtubule functions in *Drosophila* spermatogenesis require the β2-tubulin isotype-specific carboxy terminus. Devel. Biol. 1993; 158:213-227.
147. Kierszenbaum, A., Abdullah, M., Ueda, H. and Tres, L.L. Antibodies to rat sperm tail polypeptides recognize Sertoli cell secretory proteins. Molec. Cell. Biochem. 1989; 85:171-179.
148. Pisano, C., Bonascorsi, S. and Gatti, M. The kl-3 loop of the Y chromosome of *Drosophila melanogaster* binds a tektin-like protein. Genetics 1993; 133:569-579.
149. Russell, S.R.H. and Kaiser, K. *Drosophila melanogaster* male germ line-specific transcripts with autosomal and Y-linked genes. Genetics 1993; 134:293-308.
150. Li, S.S.L., O'Brien, D.A., Hou, E.W., Versola, J., Rockett, D.L. and Eddy, M. Differential activity and synthesis of lactate dehydrogenase isozymes A (muscle), B (heart) and C (testis) in mouse spermatogenic cells. Biol. Reprod. 1989; 44:173-180.
151. Weiben, E.D. Regulation of synthesis of lactate dehydrogenase X during spermatogenesis in mouse. J. Cell Biol. 1981; 88:492-498.
152. Alcivar, A.A., Trasler, J.M., Hake, L.E., Salehi-Ashtiani, K., Goldberg, E. and Hecht, N.B. DNA methylation and expression of genes coding for lactate dehydrogenase A and C during rodent spermatogenesis. Biol. Reprod. 1991; 44:527-535.
153. Wheat, C.T.E., Hintz, M. Goldberg, E. and Margoliash, E. Analyses of stage specific multiple forms of lactate dehydrogenase and of cytochrome

C during spermatogenesis in the mouse. Differentiation 1977; 9:37-41.
154. Salehi-Ashtiani, K. and Goldberg, E. Differences in regulation of testis specific lactate dehydrogenase in rat and mouse occur at multiple levels. Molec. Reprod. Devel. 1993; 35:1-7.
155. McCarrey, J.R., Berg, W.M., Paragiodakis, S.J., Zhang, P.L., Dilworth, D.D., Arnold, B.L. and Rossi, J.J. Differential transcription of Pgk genes during spermatogenesis in the mouse. Devel. Biol. 1992; 154:160-168.
156. Erickson, R.P., Kramer, J.M., Rittenhouse, J. and Salkeld, A. Quantitation of mRNAs during spermatogenesis: protamine-like histone and phosphoglycerate kinase-2 mRNAs increase after meiosis. Proc. Natl. Acad. Sci. USA 1980; 77:6086-6090.
157. Mori, C., Welch, J.E., Fulcher, K.D., O'Brien, D.A. and Eddy, E.M. Unique hexokinase messenger ribonucleic acids lacking the porin-binding domain are developmentally expressed in mouse spermatogenic cells. Biol. Reprod. 1993; 49:191-203.
158. Nanogaki, T., Noda, Y., Narimoto, K., Shiotani, M., Mori, T., Matsuda, T. and Yoshida, O. Localization of copper zinc-superoxide dismutase in the human male genital organs. Hum. Reprod. 1992; 7:81-85.
159. Jow, W.W., Schlegel, P.N., Cichon, Z., Phillips, D., Goldstein, M. and Bardin, C.W. Identification and localization of copper-zinc superoxide dismutase gene expression in rat testicular development. J. Androl. 1993; 14:439-447.
160. Shannon, M. and Handel, M.A. Expression of the Hrpt gene during spermatogenesis: Implications for sex-linked chromosome inactivation. Biol. Reprod. 1993; 49:770-778.
161. Schmitt, M.C. and Ong, D.E. Expression of cellular retinol-binding protein and lecithin-retinol acyltransferase in developing rat testis. Biol. Reprod. 1993; 49:972-979.
162. Wee, S. and Grogan, W.M. Testicular temperature-labile cholersteryl ester hydrolase. J. Biol. Chem. 1993; 268:8158-8163.
163. Sen, I., Kasturi, S. Jabbar, M.A. and Sen, G.C. Mutations in two specific residues of testicular angiotensin-converting enzyme change after its catalytic properties. J. Biol. Chem. 1993; 268:25748-25754.
164. Cheng, Y.-N.P., Ehlers, M.R.W. and Riordan, J.F. The functional role of tyrosine-200 in human angiotensin-converting enzyme. Biochem. Biophys. Res. Comm. 1992; 184:306-309.
165. Olofsson, B., Chardin, P., Touchot, N., Zahraoul, A. and Tavitian, A. Expression of the *ras*-related *ralA*, *rho12* and *rab* genes in adult mouse tissues. Oncogene 1988; 3:231-234.
166. Oyen, O., Myklebust, F., Scott, J.D., Cadd, J.D., McKnight, G.S., Hansson, V. and Jahnsen, T. Subunits of cyclic adenosine 3',5'-monophosphate-dependent protein kinase show differential and distinct expression patterns during germ cell differentiation: Alternative polyadenylation in germ cells gives rise to unique smaller-sized mRNA species. Biol. Reprod. 1990; 43:46-54.
167. Massa, J.S., Fellows, R.E. and Maurer, R.A. Rat RIβ isoform of type I regulatory subunit of cyclic adenosine monophosphate-dependent protein kinase: cRNA sequence analysis, mRNA tissue specificity, and rat/mouse differences in expression in testis. Molec. Reprod. Devel. 1990; 26:129-133.
168. Adamo, S., Conti, M., Geremia, R. and Monesi, V. Particulate and soluble

adenylate cyclase activities of mouse male germ cells. Biochem. Biophys. Res. Comm. 1980; 97:607-613.
169. Conti, M., Jin, S.-L.C., Monaco, L., Repaske, D.R. and Swinnen, J.V. Hormonal regulation of cyclic nucleotide phosphodiesterases. Endocr. Rev. 1991; 12:218-234.
170. Swinnen, J.V., Joseph, D.R. and Coti, M. Molecular cloning of rat homologues of the *Drosophila melanogaster dunce* cAMP phosphodiesterase: evidence for a family of genes. Proc. Natl. Acad. Sci. USA 1989; 86:5325-5329.
171. Mischak, H., Goodnight, J., Henderson, D.W., Osada, S., Ohno, S. and Mushinksi, J.F. Unique expression pattern of protein kinase C-θ: High mRNA levels in normal mouse testes and in T-lymphocytic cells and neoplasms. FEBS Lett 1993; 326:51-55.
172. Jinno, A., Tanaka, K., Matsushime, H., Haneji, T. and Shibuya, M. Testis-specific *Mak* protein kinase is expressed specifically in the meiotic phase in spermatogenesis and is associated with a 210-kilodalton cellular phosphoprotein. Molec. Cell. Biol. 1993; 13:4146-4156.
173. Wang, Z.Q. and Kim, K.H. Retinol differentially regulates male germ cell-asociated kinase *mak* messenger ribonucleic acid expression during spermatogenesis. Biol. Reprod. 1993; 49:951-964.
174. Letwin, K., Mizzen, L., Motro, B., Ben-David, Y. Berstein, A. and Pawson, T. A mammalian dual specificity protein kinase, *Nek1,* is related to the NIMA cell cycle regulator and is highly expressed in meiotic germ cells. EMBO J. 1992; 11:3521-3531.
175. Muramatsu, T., Giri, P.R., Higuchi, S. and Kincaid, R.L. Molecular cloning of a calmodulin-dependent phosphatase from murine testis: Identification of a developmentally expressed nonneural isozyme. Proc. Natl. Acad. Sci. USA 1992; 89:529-533.
176. Tash, J.S., Krinks, M., Patel, J., Means, R.L., Klee, C.B. and Means, A.R. Identification, characterization, and functional correlation of a calmodulin-dependent protein phosphatase in sperm. J. Cell Biol. 1988; 106:1625-1633.
177. Mukai, H., Chang, C.-D., Tanaka, H., Ito, A., Kuno, T. and Tanaka, C. cDNA cloning of a novel testis-specific calcineurin B-like protein. Biochem. Biophys. Res. Comm. 1991; 179:1325-1330.
178. Nishio, H., Matsui, H., Moia, L.J.M.P., Taketa, S., Miyamoto, K., Tokuda, M., Itano, T., Nakahara, S. and Hatase, O. The evidence for post-meiotic expression of a testis specific isoform of a regulatory subunit of calcineurin using a monoclonal antibody. Biochem. Biophys. Res. Comm. 1992; 187:828-831.
179. Hatano, Y., Shima, H., Haneji, T., Miura, A., Sugimura, R. and Nagao, M. Expression of PP2A B regulatory subunit β isotype in rat testis. FEBS 1993; 324:71-75.
180. Shima, H., Haneji, T., Hatano, Y., Kasugai, I., Sugimura, T. and Nagao, M. Protein phosphatase 1γ2 is associated with nuclei of mitotic cells in rat testis. Biochem. Biophys. Res. Comm. 1993; 194:930-937.
181. Sasaki, K., Kitagawa, Y., Shima, H., Irion, S., Sugimura, T. and Nagao, M. Production of a shorter mRNA for protein phosphatase 2Aβ by alternative polyA addition. Biochem. Biophys. Res. Comm. 1990; 170:169-175.
182. Mezquita, J., Lopez-Ibor, B., Pau, M. and Mezquita, C. Intron and

intronless transcription of chicken polyubiquitin gene UbII. FEBS Lett. 1993; 319:244-248.

183. Torii, S., Yamagishi, Y., Murakami, K. and Nakayama, K. Localization of *Kex2*-like processing endoproteases, furin and PC4, within mouse testis by in situ hybridization. FEBS 1993; 316:12-16.
184. Nakayama, K., Kim, W.-S., Torii, S., Hosaka, M., Nakagawa, T., Ikemizu, J., Baba, T. and Murakami, K. Identification of the fourth member of the mammalian endoprotease family homologous to the yeast *Kex2* protease. J. Biol. Chem. 1992; 267:5897-5900.
185. Vogl, A.W. Distribution and function of organized concentrations of actin filaments in mammalian spermatogenic cells and Sertoli cells. Int. Rev. Cytol. 1989; 119:1-56.
186. Kann, M.L., Feinberg, J., Rainteau, D., Dadoune, J.P., Weinman, S. and Fouquet, J.P. Localization of calmodulin in perinuclear structures of spermatids and spermatozoa: a comparison of six mammalian species. Anat. Rec. 1991; 230:481-488.
187. Kann, M.L., Pradel, L.A. and Fouquet, J.P. Spectrin and ankyrin like proteins in spermatids and spermatozoa of the hamster and some other mammals. Reprod. Nutr. Dev. 1993; 33:51-61.
188. Boom, J.D.G. and Smith, M.J. Molecular analyses of gene expression during sea star spermatogenesis. J. Exp. Zool. 1989; 250:312-320.
189. Stewart, M., King, K.L. and Roberts, T.M. Crystallization of the motile major sperm protein MSP of the nematode *Ascaris suum*. J. Mol. Biol. 1993; 232:298-300.
190. Varkey, J.P., Jansma, P.L., Minniti, A.N. and Ward, S. The *Caenorhabditis elegans spe-6* gene is required for major sperm protein assembly and shows second site non-complementation with an unlinked deficiency. Genetics 1993; 133:79-86.
191. Moriya, M., Katagiri, C. and Yagi, K. Immuno-electron microscopic localization of calmodulin and calmodulin-binding proteins in the mouse germ cells during spermatogenesis and maturation. Cell Tissue Res. 1993; 271:441-451.
192. Inpanbutr, N. and Taylor, A.N. Expression of calbindin-D28k in developing and growing chick testes. Histochem. 1992; 97:335-339.
193. Gruppi, C.M. and Wolgemuth, D.J. HSP86 and HSP84 exhibit cellular specificity of expression and co-precipitate with an HSP70 family member in the murine testis. Devel. Genet. 1993; 14:119-126.
194. Matsumoto, M., Jimbo, M. and Fujimoto, H. Expresion of a *hsp70*-related gene in mouse spermatid. In: Baccetti, B. ed. Comparative Spermatology 20 Years After. NY: Raven, 1991:743-746.
195. Matsumoto, M., Kurata, S., Fujimoto, H. and Hoshi, M. Haploid specific activations of protamine-1 and hsc70t genes in mouse spermatogenesis. Biochem. Biophys. Acta 1993; 1174:274-278.
196. Zakeri, Z.F., Wolgemuth, D.J. and Hunt, C.R. Identification and sequence analysis of a new member of the mouse HSP70 gene family and characterization of its unique cellular and developmental pattern of expression in the male germ line. Mol. Cell Biol. 1988; 8:2925-2932.
197. Krawczyk, Z., Mali, P. and Parvinen, M. Expression of a testis-specific *HSP70* gene-related RNA in defined stages of rat seminiferous epithelium. J. Cell Biol. 1988; 107:1317-1323.

198. Allen, R.L., O'Brien, D.A. and Eddy, E.M. A novel hsp70-like protein P70 is present in mouse spermatogenic cells. Mol. Cell. Biol. 1988; 8:828-832.
199. Allen, R.L., O'Brien, D., Jones, C.C., Rockett, D.L. and Eddy, E.M. Expression of heat shock proteins by isolated mouse spermatogenic cells. Mol. Cell. Biol. 1988; 8:3260-3266.
200. O'Rand, M.G. and Romrell, L.J. Appearance of regional surface autoantigens during spermatogenesis: comparison of anti-testis and anti-sperm autoantisera. Devel. Biol. 1980; 75:431-441.
201. Koshimizu, U., Watanabe, D., Sawada, K. and Nishimune, Y. A novel stage-specific differentiation antigen is expressed on mouse testicular germ cells during early meiotic prophase. Biol. Reprod. 1993; 49:875-884.
202. O'Brien, D.A. and Millette, C.F. Immunochemical identification of multiple cell surface antigens appearing during specific stages of mouse spermatogenesis. Gamete Res. 1986; 13:199-211.
203. Cowan, A.E. and Myles, D.G. Biogenesis of surface domains during spermiogenesis in the guinea pig. Devel. Biol. 1993; 155: 124-133.
204. Phelps, B.M. and Myles, D.G. The guinea pig sperm plasma membrane protein, PH-20, reaches the surface via two transport pathways and becomes localized to a domain after an initial uniform distribution. Devel. Biol. 1987; 123:63-72.
205. Phelps, B.M., Koppel, D.E., Primakaoff, P. and Myles, D.G. Evidence that proteolysis of the surface is an initial step in the mechanism of formation of sperm cell surface domains. J. Cell Biol. 1990; 111:1839-1847.
206. Cowan, A.E., Myles, D.G. and Koppel, D.E. Lateral diffusion of the PH-20 protein on guinea pig sperm: evidence that barriers to diffusion maintain plasma membrane domains in mammalian sperm. J. Cell Biol. 1987; 104:917-923.
207. Blobel, C.P., Myles, D.G., Primakoff, P. and White, J.M. Proteolytic processing of a protein involved in sperm-egg fusion correlates with acquisition of fertilization competence. J. Cell Biol. 1990; 111:69-77.
208. Nishioka, D., Trimmer, J.S., Poccia, D. and Vacquier, V.D. Changing localizations of site-specific surface antigens during sea urchin spermiogenesis. Exp. Cell Res. 1987; 173:606-616.
209. Trimmer, J.S., Trowbridge, I.S. and Vacquier, V.D. Monoclonal antibody to a membrane glycoprotein inhibits the acrosome reaction and associated Ca^{2+} and H^{+} fluxes of sea urchin sperm. Cell 1985; 40:697-703.
210. Geremia, R., Sorrentino, V. and Rossi, P. Oncogene expression in male germ cells at various stages of differentiation. In: Baccetti, B. ed. Comparative Spermatology 20 Years After. NY: Raven, 1991:723-726.
211. Winer, M. and Wolgemuth, D.J. Patterns of expression and potential functions of proto-oncogenes during mammalian spermatogenesis. In: deKretser, D. ed. In, Molecular Biology of the Male Reproductive System. NY: Acad.Press, 1993:143-180.
212. Muller, R., Slamon, D.J., Tremblay, J.M., Cline, M.J. and Verma, I.M. Differential expression of cellular oncogenes during pre- and postnatal development of the mouse. Nature 1982; 299:640-644.
213. Ponzetto, C. and Wolgemuth, D.J. Haploid expression of a unique *c-abl* transcript in the mouse male germ line. Mol. Cell. Biol. 1985; 5:1791-1795.

213a. Oppi, C., Shore, S.K. and Reffy, E.P. Nucleotide sequence of testis-derived *c-abl* cDNAs: Implications for testis-specific transcription and *abl* oncogene activation. Proc. Natl. Acad. Sci. USA 1987; 84:8200-8204.

214. Zakeri, Z.F., Wolgemuth, D.J. and Hunt, C.R. Identification and sequence analysis of a new member of the mouse *HSP70* gene family and characterization of its unique cellular and developmental pattern of expression in the male germ line. Mol. Cell Biol. 1988; 8:2925-2932.

215. Meijer, D. Hermans, A., von Lindern, M. van Agthoven, T., deKlein, A., Mackenbach, P., Grootegoed, A., Talarico, D., Della Valle, G. and Grosveld, G. Molecular characterization of the testis specific *c-abl* mRNA in mouse. EMBO J. 1987; 6:6041-6048.

216. Iwaoki, Y., Matsuda, H., Mutter, G.L., Watrin, F. and Wolgemuth, D.J. Differential expression of the proto-oncogenes *c-abl* and *c-mos* in developing mouse germ cells. Exp. Cell Res. 1993; 206:212-219.

217. Scheartzberg, P.L., Stall, A.M., Hardin, J.D., Bowdish, K.S., Humaran, T., Boast, S., Harbison, M.L., Robertson, E.J. and Goff, S.P. Mice homozygous for the *abl ml* mutation show poor viability and depletion of selected B and T cell populations. Cell 1991; 65:1165-1175.

218. Tybulewicz, V.L.J., Crawford, C.E., Jackson, P.K., Bronson, R.T. and Mulligan, R.C. Neonatal lethality and lymphopenia in mice with a homozygous disruption of the *c-abl* proto-oncogene. Cell 1991; 65: 1153-1163.

219. Fischman, K., Edman, J.C., Shackleford, G.M., Turner, J.A., Rutter, W.J. and Nir, U. A murine *fer* testis-specific transcript *ferT* encodes a truncated *fer* protein. Molec. Cell. Biol. 1990; 10:146-153.

220. Sorrentino, V., McKinney, M.D., Giorgi, M., Geremia, R. and Fleissner, E. Expression of cellular proto-oncogenes in the mouse male germ-line: A distinctive 2.4 kilobase *pim-1* transcript is expressed in haploid post-meiotic cells. Proc. Natl. Acad. Sci. USA 1988; 85:2191-2195.

221. Wingett, D., Reeves, R. and Magnuson, N.S. Characterization of the testis-specific *pim-1* transcript in rat. Nucl. Acids Res. 1992; 20:3183-3189.

222. Goldman, D.S., Kiessling, A.A., Millette, C.F. and Cooper, G.M. Expression of *c-mos* RNA in germ cells of male and female mice. Proc. Natl. Acad. Sci. USA 1987; 84:4509-4513.

223. van der Hoorn, F.A., Spiegel, J.E., Maylie-Pfenninger, M.-F. and Nordeen, S.K. A 43 kD *c-mos* protein is only expressed before meiosis during rat spermatogenesis. Oncogene 1991; 6:929-932.

224. van der Hoorn, F.A. Identification of the testis *c-mos* promoter: Specific activity in seminiferous tubule-derived extract and binding of a testis-specific nuclear factor. Oncogene 1992; 7:1093-1099.

225. Wadewitz, A.G., Winer, M.A. and Wolgemuth, D.J. Developmental and cell lineage specificity of *raf* family gene expression in mouse testis. Oncogene 1993; 8:1055-1062.

226. Kyriakis, J.M., App, H., Zhang, X., Banergee, P., Brautigan, D.L., Rapp, U.R. and Avruch, J. *Raf-1* activates MAP kinase-kinase. Nature 1992; 358:417-421.

227. Persson, H., Ayer-LeLievre, C., Soder, O., Villar, M.J., Metsis, M., Olson, L., Ritzen, E.M. and Hokfelt, T. Expression of β-nerve growth factor receptor mRNA in Sertoli cells downregulated by testosterone. Science 1990; 247:704-707.

228. Lamsan-Casalotti, S., Onada, M., Papadopoulos, V. and Dym, M. Developmental expression of GTP binding proteins in rat testis. J. Reprod. Fertil. 1993; 99:487-495.
229. Leung, T., How, B.E., Manser, E. and Lim, L. Germ cell β-chimaerin, a new GTPase-activating protein for *p21rac*, is specifically expressed during the acrosomal assembly stage in rat testis. J. Biol. Chem. 1993; 268:3813-3816.
230. Agnel, M., Roder, L., Vola, C. and Griffin-Shea, R. A *Drosophila rotund* transcript expressed during spermatogenesis and imaginal disc morphogenesis encodes a protein which is similar to human *Rac* GTPase-activating *rac*GAP proteins. Mol. Cell. Biol. 1992; 12:5111-5122.
231. Shackleford, G.M. and Varmus, H.E. Expression of the proto-oncogene *int-1* is restricted to postmeiotic male germ cells and the neural tube of mid-gestational embryos. Cell 1987; 50:89-95.
232. Erickson, R.P., Lai, L.-W. and Grimes, J. Creating a conditional mutation of *Wnt-1* by antisense transgenesis provides evidence that *Wnt-1* is not essential for spermatogenesis. Devel. Genetics 1993; 14:274-281.
233. Gavin, B.J., McMahon, J.A. and McMahon, A.P. Expression of multiple novel *Wnt-1/int-1*-related genes during fetal and adult mouse development. Genes Dev. 1990; 4:2319-2332.
234. Wolfes, H., Kogawa, K., Millette, C.F. and Cooper, G.M. Specific expression of nuclear proto-oncogenes before entry into meiotic prophase of spermatogenesis. Science 1989; 245:740-743.
235. Hirai, S.-I., Ryseck, R.-P., Mechta, F., Bravo, R. and Yaniv, M. Characterization of *junD:* a new member of the *jun* proto-oncogene family. EMBO J. 1989; 8:1433-1439.
236. Alcivar, A.A., Hake, L.E., Hardy, M.P. and Hecht, N.B. Increased levels of *junB* and *c-jun* mRNAs in male germ cells following testicular cell dissociation. J. Biol. Chem. 1990; 266:20160-20165.
237. Alcivar, A.A., Hake, L.E., Kwon, Y.K. and Hecht, N.B. *junD* mRNA expression differs from *c-jun* and *junB* mRNA expression during male germinal cell differentiation. Mol. Reprod. Devel. 1991; 30:187-193.
238. Sleeman,J.P. *Xenopus A-myb* is expressed during early spermatogenesis. Oncogene 1993; 8:1931-1941.
239. Kumar, G., Patel, D. and Naz, R.K. *c-MYC* mRNA is present in human sperm cells. Cell. Molec. Biol. Res. 1993; 39:111-117.
240. Naz, R.K., Ahmad, K. and Kumar, G. Presence and role of *c-myc* proto-oncogene product in mammalian sperm cell function. Biol. Reprod. 1991; 44:842-850.
241. Walker, C.W., Boom, J.D.G. and Marsh, A.G. First non-vertebrate member of the *myc* gene family is seasonally expressed in an invertebrate testis. Oncogene 1992; 7:2007-2012.
242. Bhat, N.K., Fisher, R.J., Fujiwara, S., Ascioine, R. and Papas, T.S. Temporal and tissue-specific expression of mouse *ets* genes. Proc. Natl. Acad. Sci. USA 1987; 84:3161-3165.
243. Rao, V.N., Huebner, K., Isobe, M., Ar-Rushdi, A., Croce, C.M., Shyam, E. and Reddy, P. *elk*-tissue specific *ets*-related genes on chromosomes X and 14 near translocation break points. Science 1989; 244:66-70.
244. Benbrook, D. and Pfahl, M. A novel thyroid hormone receptor encoded by a cDNA clone from a human testis library. Science 1987; 238:788-791.

245. Koenig, T.J., Lazar, M.A., Hodin, R.A. Brent, G.A., Larsen, P.R., Chin, W.W. and Moore, D. Inhibition of thyroid hormone action by a non-hormone binding *c-erbA* protein generated by alternative RNA splicing. Nature 1989; 337:659-661.
246. Lazar, M.A., Hodin, R.A. and Chin, W.W. Human carboxyl-terminal variant of α-type *c-erbA* inhibits transactivation by thyroid hormone receptors without binding thyroid hormone. Proc. Natl. Acad. Sci. USA 1989; 86:7771-7774.
247. Bernards, R., Schakleford, G.M., Gerber, M.R., Horowitz, J.M. Friend, D.H., Schartl, M., Bogenmann, E., Rapaport, J.M., McGee, T., Dryja, T.P. and Weinberg, R.A. Structure and expression of the murine retinoblastoma gene and characterization of its encoded protein. Proc. Natl. Acad. Sci. USA 1989; 86:6474-6478.
248. Partridge, J.F. and LaThangue, N.B. A developmentally regulated and tissue dependent transcription factor complexes with the retinoblastoma gene product. EMBO J. 1991; 10:3819-3827.
249. Han, K.A. and Kulesz-Martin, M.F. Alternatively spliced p53 RNA in transformed and normal cells of different tissue types. Nucl. Acids Res. 1992; 20:1979-1981.
250. Schwartz, D., Goldfinger, N. and Rotter, V. Expression of p53 protein in spermatogenesis is confined to the tetraploid pachytene primary spermatocytes. Oncogene 1993; 8:1487-1494.
251. Almon, E., Goldfinger, N., Kapon, A., Schwartz, D., Levine, A.J. and Rotter, V. Testicular tissue-specific expression of the p53 suppressor gene. Devel. Biol. 1993; 156:107-116.
252. Rotter, V., Schwartz, D., Almon, E., Goldfinger, N., Kapon, A., Meshorer, A., Donehower, L.A. and Levine, A.J. Mice with reduced levels of p53 protein exhibit the testicular giant-cell degenerative syndrome. Proc. Natl. Acad. Sci. USA 1993; 90:9075-9079.
253. Wolgemuth, D.J., Viviano, C.M. and Watrin, F. Expression of homeobox genes during spermatogenesis. Ann. N.Y. Acad. Sci. 1991; 637:300-312.
254. Wolgemuth, D.J. Aspects of homeobox gene expression in the male mammalian germ line. In: Baccetti, B. ed. Comparative Spermatology 20 Years After. NY: Raven, 1991:751-755.
255. Wolgemuth, D.J., Engelmeyer, E., Duggal, R.N., Gizang-Ginsberg, E., Mutter, G.L., Ponzetto, C., Viviano, C.M. and Zakeri, Z.F. Isolation of a mouse cDNA coding for a developmentally regulated testis-specific transcript containing homeobox homology. EMBO J. 1986; 5:1229-1235.
256. Behringer, R.R., Crotty, D.A., Tennyson, V.M., Brinster, R.L., Palmiter, R.D. and Wolgemuth, D.J. Sequences 5' of the homeobox of the *Hox-1.4* gene direct tissue-specific expression of lacZ during mouse development. Development 1993; 117:823-833.
257. Colberg-Poley, A.M., Voss, S.D., Chowdhury, K. and Gruss, P. Structural analysis of murine genes containing homeobox sequences and their expression in embryonal carcinoma cells. Nature 1985; 314:713-718.
258. Colberg-Polwy, A.M., Voss, S.D., Chowdhury, K., Stewart, C., Wagner, E.F. and Gruss, P. Clustered homeoboxes are differentially expressed during murine development. Cell 1985; 43:39-45.
259. Odenwald, W.F., Taylor, C.F., Palmer-Hill, F.J., Friedrich, V. Tani, M. and Lazzarini, R.A. Expression of a homeo domain protein in non con-

tact-inhibited cultured cells and postmitotic neurones. Genes and Devel. 1987; 1:482-496.
260. Baron, A., Featherstone, M.S., Hill, R.E., Hall, A., Galliot, B. and Duboule, D. A mouse homeo-box containing gene member of the *Hox-1* complex. EMBO J. 1987; 6:2977-2986.
261. Graham, A., Papalopulu, N., Lorimer, J., McVey, H., Tuddenham, E.G.D. and Krumlaue, R. Characterization of a murine homeo box gene, *Hox-2.6,* related to the *Drosophila* deformed gene. Genes Devel. 1988; 2:1424-1438.
262. Featherstone, M.S., Baron, A., Gaunt, S., Mattei, M.-G. and Duboule, D. *Hox-5.1* defines a homeobox-containing locus on mouse chromosome 2. Proc. Natl. Acad. Sci. USA 1988; 85: 4760-4764.
263. Watrin, F. and Wolgemuth, D.J. Conservation and divergence of patterns of expression and lineage-specific transcripts in orthologues and paralogues of the mouse *Hox-1.4* gene. Devel. Biol. 1993; 156:136-145.
264. Chauvrier, P., Vesque, C., Gaillot, B., Vigneron, M., Dolle, P., Duboule, D. and Charnay, P. The segment-specificity gene *Krox-20* encodes a transcription factor with binding sites in the promoter of the Hox-1.4 gene. EMBO J. 1990; 9:1209-1218.
265. Hatzopoulos, A.K., Stoykova, A.S., Erselius, J.R., Goulding, M., Neuman, T. and Gruss, P. Structure and expression of the mouse *Oct2a* and *Oct2b,* two differentially spliced products of the same gene. Development 1990; 109:349-362.
266. Goldsborough, A., Ashworth, A. and Willison, K. Cloning and sequencing of POU-boxes expressed in mouse testis. Nucl. Acids Res. 1990; 18:1634.
267. Rosner, M.H., Vigano, M.A., Ozato, K., Timmons, P.M., Poirier, F., Rigby, P.W.J. and Staudt, L.M. A POU-domain transcription factor in early stem cells and germ cells of the mammalian embryo. Nature 1990; 345:686-692.
268. Scholer, H.R. Octamania: The POU factors in murine development. Trends Genet. 1991; 7:323-329.
269. Scholer, H.R., Hatzopoulos, A.K., Balling, R., Suzuki, N. and Gruss, P. A family of octamer-specific proteins present during mouse embryogenesis: Evidence for germline-specific expression of an Oct factor. EMBO J. 1989; 8:2543-2550.
270. Stoykova, A.S., Sterrer, S., Erselius, J.R., Hatzopoulos, A.K. and Gruss, P. *Mini-Oct* and *Oct-2c:* Two novel, functionally diverse murine Oct-2 gene products are differentially expressed in the CNS. Neuron 1992; 8:541-558.
271. Suzuki, N., Rohdewohld, H., Neuman, T., Gruss, P. and Scholer, H.R. *Oct-6:* A POU transcription factor expressed in embyonal stem cells and in the developing brain. EMBO J. 1990; 9:3723-3732.
272. He, X., Treacy, M.N., Simmons, D.M., Ingraham, H.A., Swanson, L.W. and Rosenfeld, M.G. Expression of a large family of POU-domain regulatory genes in mammalian brain development. Nature 1989; 340:35-41.
273. Mardon, G. and Page, D.C. The sex-determining region of the mouse Y chromosome encodes a protein with a highly acidic domain and 13 zinc fingers. Cell 1989; 56:765-770.
274. Ashworth, A., Skene, B., Swift, S. and Lovell-Badge, R. Zfa is an ex-

pressed retroposon derived from an alternative transcript of the *Zfx* gene. EMBO J. 1990; 9:1529-1534.

275. Koopman, P. Ashworth, A. and Lovell-Badge, R. The *ZFY* gene family in humans and mice. Trends Genet. 1991; 7:132-136.

276. Nagamine, C.M., Chan, K., Hake, L.E. and Lau, Y.F.C. The two candidate testis-determining Y genes *Zfy-1* and *Zfy-2* are differentially expressed in fetal and adult mouse tissues. Genes Dev. 1990; 4:63-74.

276a Denny, P. and Ashworth, A. A zinc finger protein-encoding gene expressed in the post-meiotic phase of spermatogenesis. Gene 1991; 106:221-227.

277. Cunliffe, V., Koopman, P. McLaren, A. and Trowsdale, J. A mouse zinc finger gene which is transiently expressed during spermatogenesis. EMBO J. 1990; 9:197-205.

278. Cunliffe, V., Williams, S. and Trowsdale, J. Genomic analysis of a mouse zinc finger gene, *Zfp-35,* that is up-regulated during spermatogenesis. Genomics 1990; 8:331-339.

279. Burke, P.S. and Wolgemuth, D.J. *Zfp-37,* a new murine zinc finger encoding gene, is expressed in a developmentally regulated pattern in the male germ line. Nucl. Acids Res. 1992; 20:2827-2834.

280. Chowdhury, K., Goulding, M., Walther, C., Imai, K. and Fickenscher, H. The ubiquitous transactivator *Zfp-38* is upregulated during spermatogenesis with differential transcription. Mech. Devel. 1992; 39:129-142.

281. Rogers, M.B., Hosler, B.A. and Gudas, L.J. Specific expression of a retinoic acid-regulated, zinc finger gene, *Rex-1,* in preimplantation embryos, trophoblast and spermatocytes. Development 1991; 113:815-824.

282. Ernoult-Lange, M., Kress, M. and Hamer, D. A gene that encodes a protein consisting solely of zinc finger domains is preferentially expressed in transformed mouse cells. Mol. Cell. Biol. 1990; 10:418-421.

283. Chauvrier, P. Lemaire, P., Relevant, O., Bravo, R. and Charnay, P. Characterization of a mouse multigene family that encodes zinc finger structures. Mol. Cell. Biol. 1988; 8:1319-1326.

284. Chauvrier, P., Zerial, M., Lemaire, P., Almendral, J., Bravo, R. and Charnay, P. A gene encoding a protein with zinc fingers is activated during G0/G1 transition in cultured cells. EMBO J. 1988; 7:29-35.

285. Chowdhury, K., Rohdewohld, H. and Gruss, P. Specific and ubiquitous expression of different Zn finger protein genes in the mouse. Nucl. Acids Res. 1988; 16:9995-10011.

286. Hoog, C., Schalling, M., Grunder-Brundell, E. and Daneholt, B. Analysis of a murine male germ cell-specific transcript that encodes a putative zinc finger protein. Mol. Reprod. Devel. 1991; 30:173-181.

287. Boissonneault, G. and Lau, Y.-F.C. A testis-specific gene encoding a nuclear high-mobility-group box protein located in elongating spermatids. Molec. Cell. Biol. 1993; 13:4323-4330.

288. Harley, V.R., Jackson, D.I., Hextall, P., Hawkins, J.R., Berkovitz, G.D., Sockanathan, S., Lovell-Badge, R. and Goodfellow, P.N. DNA binding activity of a recombinant *SRY* from normal males and XY females. Science 1992; 255:453-456.

289. Chapman, D.L. and Wolgemuth, D.J. Identification of a mouse B-type cyclin which exhibits developmentally regulated expression in the germ line. Molec. Reprod. Develop. 1992; 33:259-269.

290. Chapman, D.L. and Wolgemuth, D.J. Isolation of the murine cyclin B2

cDNA and characterization of the lineage and temporal specificity of expression of the B1 and B2 cyclins during oogenesis, spermatogenesis and early embryogenesis. Development 1993; 118:229-240.

291. Naz, R.K. Ahmad, K. and Kaplan, P. Involvement of cyclins and cdc2 serine/threonine protein kinase in human sperm cell function. Biol. Reprod. 1993; 48:720-728.

292. Kaipia, A., Toppari, J., Mali, P., Kangasniemi, M., Alcivar, A.A. and Hecht, N.B. Stage- and cell-specific expression of the ornithine decarboxylase gene during rat and mouse spermatogenesis. Mol. Cell Endocrinol. 1990; 73:45-52.

293. Alcivar, A.A., Hake, L.E., Mali, P., Kaipia, A., Parvinen, M. and Hecht, N.B. Developmental and differential expression of the ornithine decarboxylase gene in rodent testis. Biol. Reprod. 1989; 41:1133-1142.

294. Winer, K.X.B. and Dias, J.A. Developmental regulation of ornithine decarboxylase ODC in rat testis: Comparison of changes in ODC activity with changes in ODC mRNA levels during testicular maturation. Biol. Reprod. 1992; 46:617-622.

295. Hakovirta, H., Keiski, A., Toppari, J., Halmekyto, M., Alhonen, L., Janne, J. and Parvinen, M. Polyamines and regulation of spermatogenesis: Selective stimulation of late spermatogonia in transgenic mice overexpressing the human ornithine decarboxylase gene. Molec. Endocrinol. 1993; 7:1430-1436.

296. Alcivar, A.A., Hake, L.E. and Hecht, N.B. DNA polymerase-β and polyADPribose polymerase mRNAs are differentially expressed during the development of male germinal cells. Biol. Reprod. 1992; 46:201-207.

297. Choongkittaworn, N.M., Kim, K.H. Danner, D.B. and Griswold, M.D. Expression of prohibitin in rat seminiferous epithelium. Biol. Reprod. 1993; 49:300-310.

298. Fulcher, K.D., Welch, J.E., Davis, C.M., O'Brien, D.A. and Eddy, E.M. Characterization of a laminin receptor messenger ribonucleic acid and protein expression in mouse spermatogenic cells. Biol. Reprod. 1993; 48:674-682.

299. Radhakrishnan, R., Oke, B.O., Papadopoulos, V., DeAugustine, R.P. and Suarez-Quian, C.A. Characterization of epidermal growth factor in mouse testis. Endocrinol. 1992; 131:3091-3099.

300. Kauffmann, S.H., Wright, W.W., Okret, S., Wikstrom, C., Gustafsson, J.-A., Shaper, N.L. and Shaper, J.H. Evidence that rodent epididymal sperm contain the Mr ~94,000 glucocorticoid receptor but lack the Mr ~90,000 heat shock protein. Endocrinol. 1992; 130:3074-3084.

301. Schultz, R., Isola, J., Parvinen, M., Honkaniemi, J., Wikstrom, A.-C., Gustafsson, J.A. and Pelto-Huikko, M. Localization of the glucocorticoid receptor in testis and accessory sexual organs of male rat. Molec. Cell. Endocrinol. 1993; 95:115-120.

302. Kim, K.H. and Griswold, M.D. The regulation of retinoic acid receptor mRNA levels during spermatogenesis. Mol. Endocrinol. 1990; 4: 1679-1688.

303. Rao, S.M. and Howells, R.D. *Cis*-acting elements in the 5'-untranslated region of rat testis proenkephalin mRNA regulate translation of the precursor protein. J. Biol. Chem. 1993; 268: 22164-22169.

304. Persson, H., Rehfeld, J.F., Ericsson, A., Schalling, M., Pelto-Huikko, M.

and Hokfelt, T. Transient expression of the cholecystokinin gene in male germ cells and accumulation of the peptide in the acrosomal granule: Possible role of cholecystokinin in fertilization. Proc. Natl. Acad. Sci. USA 1989; 86:6166-6170.

305. Schalling, M., Persson, H., Pelto-Huikko, M., Odum, L., Ekman, P., Gottlieb, C., Hokfelt, T. and Ritzen, E.M. Expression and localization of gastrin messenger RNA and peptide in spermatogenic cells. J. Clin. Invest. 1990; 86:660-669.
306. Persson, H., Pelto-Huikko, M., Metsis, M., Soder, O., Brene, S., Skog, S., Hokfelt, T. and Ritzen, E.M. Expression of the neurotransmitter-synthesizing enzyme glutamic acid decarboxylase in male germ cells. Mol. Cell. Biol. 1990; 10:4701-4711.
307. Ibanez, C.F., Pelto-Huikko, M., Soder, O., Ritzen, E.M., Hersh, L.B., Hokfelt, T. and Persson, H. Expression of choline acetyl transferase mRNA in spermatogenic cells results in an accumulation of the enzyme in the postacrosomal region of mature spermatozoa. Proc. Natl. Acad. Sci. USA 1991; 88:3676-3680.
308. Kew, D., Muffly, K.E. and Kilpatrick, D.L. Proenkephalin products are stored in the sperm acrosome and may function in fertilization. Proc. Natl. Acad. Sci. USA 1990; 87:1943-1947.
309. Foo, N.-C., Carter, D., Murphy, D. and Ivell, R. Vasopressin and oxytocin gene expression in rat testis. Endocrinol. 1991; 128: 2118-2128.
310. Mizuki, N., Sarapata, D.E., Garcia-Sanz, J.A. and Kasahara, M. The mouse male germ cell-specific gene *Tpx-1:* molecular structure, mode of expression in spermatogenesis, and sequence similarity to two non-mammalian genes. Mamm. Genome 1992; 3:247-280.
311. Schmid, S.R. and Linder, P. D-E-A-D protein family of putative RNA helicases. Molec. Microbiol. 1992; 6:283-292.
312. Lemaire, L. and Heinlein, U.A.O. High-level expression in male germ cells of murine P68 helicase mRNA. Life Sci. 1993; 52:917-926.
313. Leroy, P., Alzari, P., Sassoon, D., Wolgemuth, D. and Fellous, M. The protein encoded by a murine male germ cell-specific transcript is a putative ATP-dependent RNA helicase. Cell 1989; 57:549-559.
314. Karsch-Mizrachi, I. and Haynes, S.R. The *Rb97D* gene encodes a potential RNA-binding protein required for spermatogenesis in *Drosophila.* Nucl. Acids Res. 1993; 21:2229-2235.
315. Yan, Y.C., Miao, S.Y., Zong, C., Li, Y.H. Wang, L.F. and Koide, S.S. Gene expression of human sperm component related to A4 amyloid precursor protein. Arch. Androl. 1992; 28:1-6.
316. Don, J. and Wolgemuth, D.J. Identification and characterization of the regulated pattern of expression of a novel mouse gene, *megl,* during the meiotic cell cycle. Cell Growth Diff. 1992; 3:495-505.
317. Su, H., Kozak, C.A., Veerhuis, R., Lau, Y.-F. C. and Wiberg, U. Isolation of a phylogenetically conserved and testis-specific gene using a monoclonal antibody against the serological H-Y antigen. J. Reprod. Immunol. 1992; 21:275-291.
318. Tafuri, S.R., Familari, M. and Wolffe, A.P. A mouse Y box protein, MSY1, is associated with paternal mRNA in spermatocytes. J. Biol. Chem. 1993; 268:12213-12220.
319. Kwon, Y.K., Murray, M.T. and Hecht, N.B. Proteins homologous to the

Xenopus germ-cell-specific RNA-binding proteins p54/p56 are temporally expressed in mouse male germ cells. Devel. Biol. 1993; 158:90-100.

320. Ma, K., Inglis, J.D., Sharkey, A., Bickmore, W.A., Hill, R.E., Prosser, E.J., Speed, R.M., Thompson, E.J., Jobling, M., Taylor, K., Wolfe, J., Cooke, H.J., Hargreave, T.B. and Chandley, A.C. A Y chromosome gene family with RNA-binding protein homology: Candidates for the azoospermia factor AZF controlling human spermatogenesis. Cell 1993; 75:1287-1295.
321. Zhang, J.S., Yang-Feng, T.L., Muller, U. Mohandas, T.K., de Jong, P.J. and Lau, Y.-F. C. Molecular isolation and characterization of an expressed gene from the human Y chromosome. Hum. Mol. Genet. 1992; 1:717-716.
322. Mitchell, M.J., Woods, D.R., Tucker, P.K., Opp, J.S. and Bishop, C.E. Homology of a candidate spermatogenic gene from the mouse Y chromosome to the ubiquitin-activating enzyme E1. Nature 1991; 354:483-486.
323. Burgoyne, P.S., Mahadevaiah, S,K., Sutcliffe, M.J. and Palmer, S.J. Fertility in mice requires X-Y pairing and a Y chromosomal "spermiogenesis" gene mapping to the long arm. Cell 1992; 71:391-398.

CHAPTER 3

Regulation of Germ Line-Specific Expression

INTRODUCTION

Spermatogenesis is characterized by the expression of many testis-specific genes. Tissue-specific expression is often controlled at the level of transcription and most efforts are directed at defining *cis*-elements in the DNA and *trans*-acting factors which activate or repress transcription. These transcription factors can bind to the *cis*-regulatory regions of promoters or enhancers to activate RNA synthesis. Typical promoters include TATA boxes, CAAT boxes and other upstream sequences such as hormone or heat shock response elements. Enhancer elements are often found at some distance from the gene and work in either orientation. Transcription factors are defined operationally by their ability to modulate the efficiency of transcription, but DNA-binding motifs common to many of them, such as zinc finger regions or homeodomains often provide clues to function from the sequence alone. Typically, the DNA-binding domain of a transcription factor will be distinct from the activating domain which interacts with protein elements of the transcription machinery. Batteries of genes under common control may share sets of transcription factors.

Chromatin alterations can also influence gene expression. DNA sequences can be altered by methylation and chromosomal regions can be created which physically enhance or suppress expression. Hypomethylation is often correlated with actual or potential gene activity in vertebrates, and marking by methylation may be a mechanism of imprinting parental genomes. Global heterochromatization, which makes the sperm genome incompatible with gene expression, is almost universal in the last stages of spermiogenesis.

Further controls of gene expression may operate at the level of processing and export of nuclear transcripts. In particular, alternative processing of transcripts is a common mechanism in spermatogenesis, used in creating tissue-specific or multiple transcripts from a single gene. Once mRNAs reach the cytoplasm they may be stored for long periods or immediately translated. It is quite common for testicular messages to be under translational control and many mRNAs with long half-lives have been characterized, particularly those used late in spermiogenesis after transcriptional activity has ceased. Even after their synthesis, proteins can be altered by proteolytic processing, secondary covalent modifications, multimer formation and targeting to various subcellular compartments to ensure correct temporal and spatial activities.

Spermiogenesis is unusual in that it resembles a terminal differentiation process such as in, for example, lens or erythrocyte formation, yet the sperm nucleus, potentially providing the paternal genetic contribution to the zygote, must be packaged in a form in which its genes are protected from damage and from which it can be successfully reactivated after fertilization.[1,2] Reactivation mechanisms following fertilization are beyond the scope of this book, and the mechanism of global silencing of genes during spermatogenesis remains a problem not fully explored.

In this chapter, examples of various levels of control of male germ line-specific gene expression are discussed.

POTENTIAL TRANSCRIPTION FACTORS

Transcription factors can control cell-specific and temporal patterns of gene expression. Many transcription factors can be recognized by amino acid sequence motifs which predict secondary structures of their DNA-binding domains or permit interactions of subunits facilitating DNA binding by nearby regions. Most of the main categories of these motifs such as homeoboxes, zing-finger motifs, leucine zippers, POU domains, HMG boxes and helix-loop-helix or helix-turn-helix domains have been found in testicular transcription factors.[3] Steroid hormones and retinoic acid act through receptors that recognize specific DNA consensus sequences and contain some of these motifs. Some aspects of these receptors will be discussed in chapter 5. Below are discussed molecules which may act as transcription factors.

The only member of the *Hox* homeobox family expressed exclusively in testis is *Hox 1.4*.[4] Its expression may be regulated by *Krox-20,* a zinc finger protein which recognizes two 5' sequences in the Hox 1.4 gene.[5]

A variety of zinc finger protein genes whose targets have not been established have been detected in testis (see chapter 2). One well-known zinc finger transcription factor is *SP1,* which acts in a variety of cells to regulate transcription from TATA-less genes. SP1 in spermatogenesis is apparently in excess even in spermatids.[6]

Some Y-chromosome encoded genes originally thought to be involved in primary sex determination, mouse *zfy-1* and *zfy-2* and human *ZFY,* are zinc finger proteins resembling transcription factors.[7] That these genes might have a role in the beginning stages of spermatogenesis is suggested by effects of sex reversal *Srx* mutants. When portions of the *Zfy* genes are deleted in *Srx* mice, spermatogonial cell division is inhibited.[8]

A variety of hormone receptors are transcription factors containing zinc fingers. In the male germ line, *c-erb-A* encoding a T_3 thyroid hormone receptor has been reported in humans[9] but no binding of T3 occurs in rat testis.[10] *c-erbA-α-2* formed by alternative processing may function in the rat as a constitutive repressor.[11,12] A-type retinoic acid receptors have been reported in the male germ line that may be involved in regulating spermatogenesis[13] (see chapter 5). Their transcripts increase early in spermatogenesis at Stages VII-IX.[13]

A transcription factor with both leucine zipper and helix-loop-helix motifs is the nuclear oncogene *c-myc.*[14,15] It is expressed only in spermatogonia of the mouse.[16]

Leucine zipper motifs are found in *c-jun, junB, junD,* and *c-fos.* The transcription factor AP-1 is a dimeric complex of *c-fos* and *c-jun* or related genes interacting through leucine zippers which bring together the DNA binding regions of each subunit. Three categories of expression of AP-1 mRNA and protein family members have been distinguished during spermatogenesis in the fox: continuous *(c-fos),* expression only at the beginning and at the end of spermatogenesis *(fra-1, fra-2,* and *c-jun),* or expression only at the beginning of spermatogenesis *(junB).*[17] This suggests distinct roles for various AP-1 complexes during progression of spermatogenesis.

An interesting set of leucine zipper transcription factors is the CREB family which mediates expression of genes controlled by cAMP levels. CREB proteins are nuclear targets for kinase-A. Phosphorylated CREBs bind to CREs, cAMP responsive elements which are found in promoters of glycoprotein and somatostatin genes and activate transcription. The related CREMs, cyclic-AMP-responsive element modulators, are antagonists of CREBs, repressing transcription. Alternative splicing generates several repressors and activators of cAMP responses[18] and from a single transcript alternative translation initiation codons can give transcriptional activators or repressor proteins.[19] Four different CREBs are expressed in the male germ line,[20] two of which lack the leucine zipper and may act negatively by sequestering active transcription factors. The CREB testis-specific isoforms are truncated and lack DNA-binding and nuclear translocation sequences.[21]

CREM is converted from an antagonist to activator beginning in premeiotic spermatocytes, when CREB control is lost.[22] A switch in CREM gene expression occurs by alternative splicing. The activator CREM-τ, a leucine zipper protein, has two glutamine-rich domains conferring transcriptional activation activity.[23] Accumulation of high levels of messenger is stimulated by FSH and is caused by alternative polyadenylation using the most 5' polyadenylation site thus enhancing stability of the mRNA.[22] CREM-α, -β and -γ transcripts are expressed at low levels in premeiotic cells, then turned off at meiotic prophase although continuing in Sertoli and Leydig cells. CREM-τ increases greatly and becomes the sole CREM expressed starting at meiotic prophase. The protein is restricted to round spermatids and activated by phosphorylation at Ser-117 by a kinase-A.[24] It binds to CREs in the promoters of several genes whose expression is stimulated in haploid germ cells such as protamines-1 and -2, TP 1, thymosin, somatostatin, Tpx-1a and -1b, and RT7. The in vitro transcription of RT7 is inhibited by antibodies to CREM.[24]

A gene called *Sprm-1* encodes a putative transcription factor containing a POU domain and related to the Oct3/4 transactivator.[25] POU domains consisting of POU-specific and POU homeodomains connected by a short variable region, confer high-affinity DNA binding and protein-protein interactions. *Sprm-1* is expressed for approximately 40 hr just before meiosis I in male germ cells corresponding to Stages XII and XIII (late pachytene and early diplotene). It is a candidate for a meiotic regulator of, for example, testis-specific histones which possess octamer elements. An *Oct-2d* transcript appears in spermatids.[26]

An example of the HMG box is found in the *SRY* gene involved in sex-determination, suggesting a role as a transcription factor (see Chap. 6). Potential targets for the HMG box are found in the upstream regions of the *SRY* genes themselves, suggesting a feedback regulation upon its own production.[27]

PROMOTERS

Promoter and enhancer regions have been defined for several genes expressed testis- specifically[28] such as protamines-1 and -2, transition protein TP 1, testis-specific histone TH2B,[29] histone H1t, *Hox* 1.4,[30] phosphoglycerate kinase-2, *c-mos,* and RT7.[3] Transfection of gene constructs into cultured cells has been of limited utility due to the unavailability of suitable culture systems for mammalian male germ line cells. In vitro transcription assays and construction of transgenic mice, however, have provided opportunities for analysis of regulatory elements. Common motifs involved in tissue-specific expression are beginning to emerge. Other sequences are so far unique. Some aspects of promoter organization are reviewed in ref. 28. An illuminating discussion of the use of transgenics in the study of spermatogenesis, including complications due to inappropriate expression and the effects of transgenes on fertility through expression of the reporter gene or insertional mutagenesis, can be found in ref. 3.

Among the most thoroughly studied promoters are those associated with the protamine genes. Three elements (B, D, E) are involved in regulation of protamine-1 and protamine-2 genes of mouse. Deletions of 5' elements increases transcription levels in in vitro systems.[31] Mouse protamine-1[32] and protamine-2[33] have several common sequences in the 5' flanking regions and highly conserved sequences next to the polyadenylation signal in the 3' region. Although protamine-1, protamine-2 and TP 1 are all expressed post-meiotically, there are no conserved 5' regions common to the three.[28] All known protamine genes to date have 3' region homologies.[28]

In a comparison of the promoter regions of protamine-1 from ten mammals, three conserved regions were detected.[34] The first consensus, A/C/T-C/G-A/C-C-C/T-T-C-A-C/T-A- A-T is found in all protamine-1 molecules. Its central core is present in a number of other testis-specific genes such as protamine-2, TP 1, TP 2, angiotensin converting enzyme, phosphoglycerate kinase-2, proacrosin, histone H3.3B and polyubiquitin UbII.[34] The sequence binds a *trans*-acting factor from mouse testis, Tet-1.[35] A second consensus, TGTGAGG, is 20 nucleotides upstream from the first. A third, at -113 to -132 is similar to the *c-fos* serum response element. The second and third are not found in other sperm-specific genes.

Transgenic mice were made with a mouse protamine-1 promoter linked to a human growth hormone reporter. The gene was expressed in homozygous and heterozygous individuals in essentially all late spermatids suggesting that the mRNAs can diffuse through the syncytium of late spermatids.[36] The promoter has a 113-bp region directing spermatid-specific expression.[37] The 5' region is sufficient for cell-specific expression. Tissue and cell-specific expression required a region from -40 to -465 bp of the transcription start site.[38,39] It includes a cAMP response-like element 20 bp downstream from the E element which moderately activates transcription in nuclear extracts and that binds testis-specific nuclear proteins.[35] Information in an 859 bp sequence of the mouse protamine-2 5' region is required for tissue-specific expression.[40]

The mouse histone *H1t* gene, in addition to SP1, CAAT and TATA elements, contains a testis-specific promoter region homologous to a *Drosophila* β2-tubulin element directing tissue-specific expression (see below).[41] It is found between the CCAAT and SP1 sites and binds a nuclear protein of 70-80 kDa from mature testis nuclei.[41] An 18-bp element in the promoter binds only testis-specific proteins of 13 and 30 kDa.[42] These proteins first appear in pachytene spermatocytes, consistent with their roles as potential transcriptional regulators.[43] They also bind to conserved sequences from human and monkey *H1t* genes.[44] *H1t* promoter constructs are expressed in a replication-dependent fashion in somatic cells after transfection, although *H1t* is expressed in a replication-independent way in testis.[45] The proximal promoter of the rat histone H4t gene contains two elements that bind specifically to proteins of pachytene spermatocytes.[46]

The promoters of testis histones TH2A and TH2B are very similar to their somatic H2A and H2B counterparts,[47] but these histones are expressed in a replication-independent fashion. Surprisingly, rat testicular H2A and H2B have 5' sequences associated with S-phase control and yet the histones are not linked to S phase[48] so these regions may be silenced. The *TH2B* gene has two elements, a CAAT box and an octamer ATTTGCAT which allow it to be expressed in an S-phase dependent manner when introduced into somatic cells.[49] Interacting proteins binding these elements are present at constant levels during the cell cycle suggesting regulation at the post-translational level. These elements bind repressor proteins from spermatocytes that presumably prevent cell cycle regulation.[50] TH2B is expressed in spermatocytes. Proteins from premeiotic cells bind between the TATA and transcription start sites to repress expression. How TH2B is turned off in spermatids is not known.

The basis of testis-specific histone gene expression in the sea urchin involves octamer sequences. The sea urchin testis-specific *Sp H2B* histone gene has two octamer and two CAAT motifs involved in transcriptional control.[51] A testis-specific octamer binding protein of 85 kDa[52] distinct from other tissue-specific octamer binding proteins may partially regulate transcription of the *Sp H2B* gene in spermatogenesis. A CAAT binding factor in testis is only able to bind to the *Sp H2B* promoter, and an embryonic CAAT displacement factor binds to the promoter and excludes the binding factor, suggesting it acts as a repressor of Sp H2B in the embryo.[51]

The promoter of the testis-specific cytochrome c_S transcript has sequences in common with other genes expressed post-meiotically, including several regions of homology with the mouse protamine-2 gene responsible for its transcriptional repression.[53] Little sequence similarity has been detected in the promoter regions of somatic and testis cytochrome c genes in the rat.[54]

Somatic and testicular angiotensin-converting enzymes are encoded by the same gene. ACE_t mRNA is first detected in late pachytene spermatocytes but the protein only in Step 10 spermatids.[55] The testicular form has a promoter located in an intron 7 kb inside the somatic gene.[56] Cell and stage-specific expression requires a region 91 bp upstream which includes a cyclic AMP-responsive element (CRE) interacting with a testis-specific transactivator.[57] The element is also found in other CRE-like sequences such as in the protamine[58] and RT7[59] genes.

The rat RT7 promoter has CREB sites (T sites) and a region in common with protamine-1 and protamine-2 (D element).[59] The T sites bind a 39 kDa nuclear protein from testis. A factor specific to seminiferous tubules (TTF-D) binds to the D element and a second factor which binds to a downstream element probably blocks somatic transcription.

The *c-mos* promoter has a CAAT box, AP-1 sites and sites for helix-loop-helix transcription factors.[60] An element highly homologous to a region of the RT7 promoter also binds the testis-specific factor TTF-D.[60]

The testis-specific phosphoglycerate kinase PGK-2 gene in mouse has a silencer-like *cis* element in the distal upstream region (-1401 to -685) responsible for repressing PGK-2 in somatic tissue and a positive control element in the proximal region promoting expression.[61] The latter element binds a nuclear factor called TAP-1 in the sequence 5'-GGAA-3'. A nuclear factor TIN- 1, a transcriptional inhibitor of the somatic PGK-1, binds to a similar sequence.[62] Both elements are similar to the binding element for the proto-oncogene *ets.* It is possible that the two factors are identical and will simultaneously inhibit somatic- and stimulate testis-specific gene expression. The human PGK-2 promoter contains a 327 bp tissue-specific enhancer 180 bp upstream of the core promoter directing cell-specific expression in transgenic mice.[63] A 40 bp region of the enhancer (E1 and E2) binds testis-specific transcription factors from adult testis that promote expression, and an E4 element binds a protein from somatic cells that may act as a silencer.[64]

Pyruvate dehydrogenase (Pdha-2), the testis form of the E1α subunit of pyruvate dehydrogenase, has a complex promoter with several transcription factor binding sites.[65] One site (MEP-2) can bind both testis-specific and somatic variant factors inversely, the interaction with the specific factor occurring at the time of gene activation and the second decreasing as spermatogenesis proceeds. A second site (MEP-3) interacts with a ubiquitous protein. The MEP- 2 binding sequence is very similar to the 17-bp sequence reported to be involved in positive control of PGK-2[64] suggesting a common transcription factor.

Heme oxygenase (HO) gene expression is transcriptionally controlled in mouse. It is activated in mouse spermatogonia. HO mRNA is stored in spermatocytes and its synthesis continues throughout meiosis. The gene has an upstream TRE (TPA-responsive element). The transcriptional factors *c-jun* and *c-fos* are activated in spermatogonia through c-kinase activation which is probably responsible for turning on the HO gene by interaction with the TRE in this cell type.[66]

Proenkephalin mRNAs are generated from alternative germ cell-specific promoters in rat.[67-69] There is a TATA-less alternative promoter within the first intron for the somatic transcript. In transgenic mice, genes with somatic and germ-line promoters express normally. Sequences within a 500-bp region of the germ cell promoter in rat are required for tissue-specific expression in transgenic mice.[67] The upstream somatic promoter neither increases nor decreases levels of expression. Eight potential regulatory regions have been identified several of which contain consensus sequences for binding known transcription factors such as AP2, PEA3, SP1, NF-E1-like factors, IE1.2 enhancer binding proteins and GLI-like proteins. Human proenkephalin has a single promoter for somatic and germ line transcripts[70] but is properly transcribed in transgenic mouse germ cells, suggesting that mouse cells contain nuclear regulatory factors which interact with the human gene.[71,72]

The promoter of the adult testis-specific *Hox* 1.4 was investigated in transgenic mice.[73] Deletion experiments revealed a *cis* acting element in a region from 2-4 kb upstream of the transcription site.

Rat *hst70* tissue- and development-specific gene expression is regulated by a 0.8 kb upstream region in transgenic mice.[74] The transgene transcripts accumulated in pachytene spermatocytes. The 0.8 kb region contains TATA, CAAT, octamer and Sp1 boxes, and putative estrogen and heat shock response elements, but no homology to protamine or PGK-2 promoters which would indicate tissue-specific *cis*-elements.

A 2.3 kb region of the proacrosin rat promoter 5' to the protein initiation codon is sufficient to direct appropriate cell-type expression of a reporter gene to pachytene spermatocytes and early spermatids.[75] The protein was not however packaged into the acrosome.

The *Drosophila* β2 tubulin gene is transcribed in meiotic cells but the protein continues to be synthesized in spermatids. A 14-bp promoter element is required for testis-specific transcription of the gene.[76]

ROLES OF METHYLATION

Methylation of DNA is believed to play a role in gene expression of mammalian cells. DNA methylation may directly alter transcription factor-DNA binding or may influence it more indirectly through changes in chromatin structure. Methylation of cytidine-rich regions at CpG sites of tissue-specific genes is generally reversed when the genes are expressed. If methylation inhibits expression, hypomethylation should be correlated with expression of germ line-specific genes.

Changes in methylation state of several testis-specific genes have been investigated without clear trends. Mouse transition protein TP 1 becomes hypomethylated during spermatogenesis whereas protamines-1 and-2 become more methylated.[77] Lactate dehydrogenase LDH-C and LDH-A show no changes in methylation pattern during spermatogenesis,[78] nor does TH2B.[79] LDH-C methylation was the same in testis and spleen.[78]

No differences were found in phosphoglycerate kinase PGK-2 from pachytene spermatocytes to round spermatids.[77,80] Yet when compared to somatic cells and sperm, which should be more highly methylated, sites within PGK-2 of spermatogenic cells were found to be less methylated.[80] Thus no simple correlation exists for hypomethylation of testis-specific genes during the time of their expression or compared to somatic or sperm genes, at least for the sites examined to date.

A transgene of the human testis-specific lactate dehydrogenase driven by the mouse metallothionein promoter was expressed only in testis.[81] Its repression at the transcriptional level in somatic tissue correlated with hypermethylation. In the testis, the gene was hypomethylated and developmentally regulated. A transgene for troponin I is undermethylated in testis relative to somatic tissue.[82]

A role for methylation may occur in genomic imprinting.[83] Differences in gene expression of some genes depends upon from which parent they are inherited. The pattern of imprinting can be changed when the gene passes through the germ line of the opposite sex. Examples of endogenous imprinted genes in the mouse are insulin-like growth factor IGF-II[84] and the IGF-II receptor gene,[85] the Tme gene,[86] and the H19 gene.[87] Several transgenes also show imprinting. Changes in gene expression of some imprinted genes and transgenes are correlated with changes in methylation. For a more complete discussion of methylation of endogenous and transgenes, see ref. 3.

ALTERNATIVE TRANSCRIPTS AND PROCESSING

One way of producing testis-specific transcripts is to make a different transcript from the same gene serving for the somatic transcript by alternative splicing or processing. Such mechanisms are common for postmeiotic transcripts.[88] A transcript may be altered by changing the site of initiation, splicing differentially in the coding or untranslated regions, or using various polyadenylation sites.

Alternative transcription initiation sites are used for *c-mos*,[89] cytochrome c_T,[53,90] proenkephalin,[68,91] β1,4-galactosyltransferase[92] and angiotensin-converting enzyme.[93] In the case of ACE, a 698 bp fragment containing the testis-specific promoter immediately upstream from the transcription start site was demonstrated by construction of a transgenic mouse which directed tissue-specific expression. β-Galactosyltransferase and cytochrome c_T are first expressed in round spermatids.[94] β-Galactosyltransferase mRNAs of 2.9 and 4.1 kb are produced by alternative use of 3' poly(A) signals. The 4.1 kb transcript is replaced by the 2.9 and a 3.1 kb mRNA produced by use of a haploid-specific promoter.[94] Alternative splicing in *fer* changes the protein coding region,[95] whereas in prodynorphin a region of the 5' UTR is removed.[68] A short *c-abl* transcript is produced by premature transcription termination using an alternative polyadenylation site not preceded by a consensus polyadenylation signal.[96] Four isoforms of the regulatory subunit of protein kinase A are differentially expressed during spermatogenesis, the mRNAs for the two smaller forms produced by alternative polyadenylation.[97]

TRANSLATIONAL CONTROL

Many proteins are translated in spermatids well after nuclear transcription is shut down and the chromatin becomes condensed. Their mRNAs therefore must be stable and some are stored for a considerable period before being activated. Elements in the 5'- or 3'-untranslated regions (UTRs) and the length of the poly(A) tails may control timing and stability of the messages.

In rat proenkephalin, a longer germ cell-specific cDNA has a novel 5' terminus with many short open reading frames[68] derived from a somatic gene intron. Four upstream AUGs cause the germ cell-specific transcripts to translate less efficiently in vitro which may account for why so little protein is made despite the abundance of mRNA in pachytene spermatocytes and round spermatids.[98]

Transition proteins are synthesized in mid-spermiogenesis and protamines at the end of spermiogenesis, each well after transcription ceases. Thus TP 1 mRNA has to be stored for about 3 days and protamine mRNA for about 7 days. Each undergoes shortening due to loss of some of the poly(A) tail, which is accompanied by a shift from a non-polysomal to a polysomal location.[28]

The 3' UTR of protamine-1 controls timing of expression. In a comparison of the 3' UTR from mouse protamine-1 with that of human growth hormone in otherwise identical transgenic constructs, the protamine UTR shifted translation from elongating spermatids to round spermatids,[38] although the time of transcription was identical for each. In addition the mRNA with the protamine-1 3' UTR exhibited increased stability.

The 3' untranslated region of mouse protamine-2 mRNA binds a phosphoprotein which represses its translation.[99] Two conserved sequences (Y and H elements) in the 3' region bind the 18 kDa protein from post-mitochondrial supernatant fractions of testis homogenates. The protein is maximal in round spermatids in which protamine-2 mRNA is stored as an inactive mRNP particle. Its presence in round but not elongating spermatid cytoplasm suggests that it may prevent expression. The protein needs to be phosphorylated to bind to RNA.[99,100] The Y and H sequences are found in 3' UTRs of other translationally regulated transcripts in germ cells such as protamine-1 and TP 1 in rat, mouse and human. In both they are associated with predicted stem-loop structures.

The human TP 2 transcript is expressed at very low levels compared to other mammals. It contains a deletion in the 3' UTR of an 8 bp motif GCCATCAC. This motif binds a cytosolic protein from rat testis but a mutant lacking the motif does not.[101] It is suggested that the motif confers storage stability to the TP transcript.

Several mRNAs are shortened by loss of some of the poly(A) tail during mammalian spermatogenesis including protamines-1 and -2, TP 1 and TP 2, and the mitochondrial selenoprotein.[28,102] In contrast, activation of PGK-2, and *jun* translation is accomplished without change in mRNA length.[103,104]

The shortening of the poly(A) tracts cannot be the only factor involved in translational activation of protamine, TP and selenoprotein

(MCS) mRNAs.[105] Mouse TP 1, protamine-1 and protamine-2 polysomes have more closely spaced ribosomes suggesting more rapid initiation or slower elongation. In contrast MCS ribosomes are spaced more widely and probably suffer inefficient insertion of selenocysteine at the codon UGA, normally a stop signal. TP 2 appears to be a weak competitor for a limiting initiation factor. Experiments with protamine-2 suggest active single ribosomes may be limiting in late spermatids.

Cytochrome c_T mRNA exhibits translational control. Larger mRNAs of 0.7-0.9 kb are found in polysomes in pachytene spermatocytes in addition to smaller non-polysomal 0.6-0.75 kb transcripts.[90] In round spermatids 0.7 kb mRNAs are mostly non-polysomal molecules. Two 3' sequences near the poly(A) addition signal are reminiscent of AU rich cytoplasmic polyadenylation elements (CPEs) that may mediate translational repression.[90] The longer poly(A) tails in pachytene spermatocytes may activate translation. Cytochrome c_S transcripts persist into late stages but are poorly translated.[90]

Lactate dehydrogenase C, the testis-specific form of LDH, has two mRNAs differing in poly(A) tail length (1.2 and 1.3 kb). The longer ones are associated with polysomes in pachytene spermatocytes and round spermatids.[78] The transcripts contain CPEs that may regulate their expression. LDH-A mRNAs accumulate in non-polysomal form at late stages of spermatogenesis and are not translated efficiently.[78]

Two genes translationally controlled in *Drosophila* are the mst(gl)-9m and Janus B genes. Mst(gl)-9m is transcribed premeiotically but translated post-meiotically (3 days later) under control of elements in the 5' untranslated region.[106,107] Janus B shows a similar pattern.[108] Both are reminiscent of mouse protamine-1. An 18-bp sequence in the 5' untranslated region of the *Drosophila* β2-tubulin mRNA affects mRNA levels post-meiotically by controlling message stability.[109] It is position independent.

Translationally controlled transcripts of the Mst(3)CGP family are expressed testis-specifically.[110] Two family members, Mst98Ca and Mst98Cb, exclusively expressed in the male germ line, encode proteins with a repetitive Cys-Gly-Pro motif that has homology with proteins of the outer dense fiber of mammalian sperm. The protein localizes to the *Drosophila* sperm tail.[111] Along with other members of the family, the mRNAs have a conserved 5' UTR element called TCE controlling delayed translational expression and elongation of the poly(A) tail. The 12 nt long TCE is always +28 nt from the transcription site. Its function is dependent on its correct position. Proteins binding to the TCE of in vitro transcripts of the Mst87F gene in *Drosophila* testis extracts have been demonstrated.[111] One can be directly cross-linked by ultraviolet light to the transcript. The proteins are testis-specific but conserved between *Drosophila* species. The TCE can also act as a testis-specific transcriptional regulator and specific proteins binding to the element can be found in a variety of *Drosophila* tissues.[111] It is similar to Y box proteins (see chapter 2) in exhibiting both transcriptional and translational regulation.

Another RNA binding-protein which may regulate translation in *Drosophila* is encoded by *orb*.[112] It is a predicted 99 kDa protein transcribed specifically in germ line tissue and expressed from late third instar larvae to adult.

SUMMARY

Regulation of germ line gene expression occurs at many levels. Positive and negative *trans*-acting factors interact with *cis*-regulatory regions containing testis-specific motifs. Alternative processing of transcripts provides variation from a single gene region. The timing of translational initiation and stability of mRNAs are also controlled. The importance of DNA methylation is not yet clear.

REFERENCES

1. Medvedev, Z.A. On the immortality of the germ line: genetic and biochemical mechanisms. A review. Mech. Aging and Develop. 1981; 17:331-359.
2. Poccia, D.L. and Green, G.R. Packaging and unpackaging the sea urchin genome. Trends Biochem. Sci. 1992; 17:223-227.
3. Eddy, E.M., Welch, J.E. and O'Brien, D.A. Gene expression during spermatogenesis. In: deKretser, D. ed. Molecular Biology of the Male Reproductive System. NY: Acad.Press, 1993:181-232.
4. Wolgemuth, D.J., Engelmeyer, E., Duggal, R.N., Gizang-Ginsberg, E., Mutter, G.L., Ponzetto, C., Viviano, C.M. and Zakeri, Z.F. Isolation of a mouse cDNA coding for a developmentally regulated testis-specific transcript containing homeobox homology. EMBO J. 1986; 5:1229-1235.
5. Chauvrier, P., Vesque, C., Gaillot, B., Vigneron, M., Dolle, P., Duboule, D. and Charnay, P. The segment-specificity gene *Krox-20* encodes a transcription factor with binding sites in the promoter of the *Hox-1.4* gene. EMBO J. 1990; 9:1209-1218.
6. Saffer, J.D., Jackson, S.P. and Annarella, M.B. Developmental expression of Sp1 in the mouse. Mol. Cell Biol. 1991; 11:2189-2199.
7. Mardon, G. and Page, D.C. The sex-determining region of the mouse Y chromosome encodes a protein with a highly acidic domain and 13 zinc fingers. Cell 1989; 56:765-770.
8. Koopman, P. Ashworth, A. and Lovell-Badge, R. The *ZFY* gene family in humans and mice. Trends Genet. 1991; 7:132-136.
9. Bennbrook, D. and Pfahl, M. A novel thyroid hormone receptor encoded by a cDNA clone from a human testis library. Science 1987; 238:788-791.
10. Jannini, E.A., Olivieri, M., Francavilla, S., Gulino, A., Ziparo, E. and D'Armiento M. Ontogenesis of the nuclear 3,5,3'-triiodothyronine receptor in the rat testis. Endocrinol. 1990; 126:2521-2526.
11. Koenig, T.J., Lazar, M.A., Hodin, R.A. Brent, G.A., Larsen, P.R., Chin, W.W. and Moore, D. Inhibition of thyroid hormone action by a non-hormone binding *c-erbA* protein generated by alternative RNA splicing. Nature 1989; 337:659-661.
12. Lazar, M.A., Hodin, R.A. and Chin, W.W. Human carboxyl-terminal variant of α-type *c-erbA* inhibits transactivation by thyroid hormone receptors without binding thyroid hormone. Proc. Natl. Acad. Sci. USA 1989; 86:7771-7774.

13. Kim, K.H. and Griswold, M.D. The regulation of retinoic acid receptor mRNA levels during spermatogenesis. Mol. Endocrinol. 1990; 4: 1679-1688.
14. Wolfes, H., Kogawa, K., Millette, C.F. and Cooper, G.M. Specific expression of nuclear proto-oncogenes before entry into meiotic prophase of spermatogenesis. Science 1989; 245:740-743.
15. Koji, T., Izumi, S., Tanno, M., Moriuchi, T. and Nahane, P.K. Localization in situ of *c-myc* mRNA and *c-myc* protein in adult mouse testis. Histochem. J. 1988; 20:551-557.
16. Stewart, T.A., Hecht, N.B., Hollingshead, P.G., Johnson, P.A., Leong, J.A.C. and Pitts, S.L. Haploid-specific transcription of protamine-*myc* and protamine-T-antigen fusion genes in transgenic mice. Mol. Cell. Biol. 1988; 8:1748-1755.
17. Cohen, D.R., Vandermark, S.E., McGovern, J.D. and Bradley, M.P. Transcriptional regulation in the testis: a role for transcription factor AP-1 complexes at various stges of spermatogenesis. Oncogene 1993; 8:443-455.
18. Foulkes, N.S. and Sassone-Corsi, P. More is better: activators and repressors from the same gene. Cell 1992; 68:411-414.
19. Delmas, V., Laoide, B., Masquillier, D., de Groot, R.P., Foulkes, N.S. and Sassone-Corsi, P. Alternative usage of initiation codons in CREM generates regulators with opposite functions. Proc. Natl. Acad. Sci. USA 1992; 89:4226-4230.
20. Ruppert, S., Cole, T.J., Boshart, M., Schmid, E. and Schutz, G. Multiple mRNA isoforms of the transcription activator protein CREB: Generation by alternative splicing and specific expression in primary spermatocytes. EMBO J. 1992; 11:1503-1512.
21. Waeber, G., Meyer, T.E., LeSieur, M., Hermann, H.L., Gerard, N. and Haabener, J. Developmental stage-specific expression of cyclic adenosine 3',5'-monophosphate response element-binding protein CREB during spermatogenesis involves alternative exon splicing. Molec. Endocrinol. 1991; 5:1418-1430.
22. Foulkes, N.S., Schlotter, F., Pevet, P. and Sassone-Corsi, P. Pituitary hormone FSH directs the CREM functional switch during spermatogenesis. Nature 1993; 362:264-267.
23. Foulkes, N.S., Mellstrom, B., Benusiglio, E. and Sassone-Corsi, P. Developmental switch of CREM function during spermatogenesis: from antagonist to activator. Nature 1992; 355:80-84.
24. Delmas, V., van der Hoorn, F., Mellstrom, B., Jegou, B. and Sassone-Corsi, P. Induction of CREM activator proteins in spermatids: Downstream targets and implications for haploid germ cell differentiation. Molec. Endocrinol. 1993; 7:1502-1514.
25. Andersen, B., Pearse, R.V., Schlegel, P.N., Cichon, Z., Schonemann, M.D., Bardin, C.W. and Rosenfeld, M.G. Sperm 1: A POU-domain gene transiently expressed immediately before meiosis I in the male germ cell. Proc. Natl. Acad. Sci. USA 1993; 90: 11084-11088.
26. Stoykova, A.S., Sterrer, S., Erselius, J.R., Hatzopoulos, A.K. and Gruss, P. Mini-Oct and Oct-2c: Two novel, functionally diverse murine Oct-2 gene products are differentially expressed in the CNS. Neuron 1992; 8:541-558.

27. Harley, V.R., Jackson, D.I., Hextall, P., Hawkins, J.R., Berkovitz, G.D., Sockanathan, S., Lovell-Badge, R. and Goodfellow, P.N. DNA binding activity of a recombinant SRY from normal males and XY females. Science 1992; 255:453-456.
28. Hecht, N.B. Gene expression during male germ cell development. In: Desjardins C. and Ewing, L.L. eds. Cell and Molecular Biology of the Testis. Oxford: Oxford Univ. Press, 1992:464-503.
29. Kim, Y.-J., Hwand, I., Tres, L.L., Kierszenbaum, A.L. and Chae, C.B. Molecular cloning and differential expression of somatic and testis-specific H2b histone genes during rat spermatogenesis. Devel.Biol. 1987; 124:23-34.
30. Wolgemuth, D.J., Gizang-Ginsberg, F., Ponzetto, C. and Zakeri, Z.F. Genetic control of germ cell function: developmentally regulated gene expression during gametogenesis. In: Schatten, G. and Schatten, H., eds. Molecular Biology of Fertilization. NY: Acad. Press, 1988:235-258.
31. Bunick, D., Johnson, P.A., Johnson, T.R. and Hecht, N.B. Transcription of the testis-specific mouse protamine 2 gene in a homologous in vitro transcription system. Proc. Natl. Acad. Sci. USA 1990; 87:891-895.
32. Johnson, P.A., Peschon, J.J., Yelick, P.C., Palmiter, R.D. and Hecht, N.B. Sequence homologies in the mouse protamine 1 and 2 genes. Biochim. Biophys. Acta 1988; 950:45-53.
33. Krawetz, S.A., Connor, W. and Dixon, G.W. Bovine protamine genes contain a single intron. J. Biol. Chem. 1988; 263:321-326.
34. Queralt, R. and Oliva, R. Identification of conserved potential regulatory sequences of the mammalian protamine-encoding P1 genes from ten different mammals. Gene 1993; 133:197-201.
35. Tamura, T. Makino, Y., Mikoshiba, K. and Maramatsu, M. Demonstration of a testis-specific trans-acting factor *Tet-1* in vitro that binds to the promoter of the mouse protamine 1 gene. J. Biol. Chem. 1992; 267:4327-4332.
36. Braun, R.E., Behringer, R.R., Peschon, J.J., Brinster, R.L. and Palmiter, R.D. Genetically haploid spermatids are phenotypically diploid. Nature 1989; 337:373-376.
37. Zambrowicz, B.P., Harendza, C.J., Zimmerman, J.W., Brinster, R.L. and Palmiter, R.D. Analysis of the mouse protamine 1 promoter in transgenic mice. Proc. Natl. Acad. Sci. 1993; 90:5071-5075.
38. Braun, R.F., Peschon, J.J., Behringer, R.R., Brinster, R. and Palmiter, R.D. Protamine 3'-untranslated sequences regulate temporal translational control and subcellular localization of growth hormone in spermatids of transgenic mice. Genes Devel. 1989; 3:793-802.
39. Peschon, J.J., Behringer, R.R., Brinster, R.L. and Palmiter, R.S. Spermatid-specific expression of protamine 1 in transgenic mice. Proc. Natl. Acad. Sci. USA 1987; 84:5316-5319.
40. Stewart, T.A., Hecht, N.B., Hollingshead, P.G., Johnson, P.A., Leong, J.A.C. and Pitts, S.L. Haploid-specific transcription of protamine-*myc* and protamine-T-antigen fusion genes in transgenic mice. Mol. Cell. Biol. 1988; 8:1748-1755.
41. Wolfe, S.A. and Grimes, S.R. Histone H1t: A tissue-specific model used to study transcriptional control and nuclear function during cellular differentiation. J. Cell. Biochem. 1993; 53:156-160.

42. Grimes, S.R., Wolfe, S.A. and Koppel, D.A. Tissue-specific binding of testis nuclear proteins to a sequence element within the promoter of the testis-specific histone H1t gene. Arch. Biochem. Biophys. 1992; 296:402-409.
43. Grimes, S.R., Wolfe, S.A. and Koppel, D.A. Temporal correlation between the appearance of testis-specific DNA-binding proteins and the onset of transcription of the testis-specific histone H1 gene. Exp. Cell Res. 1992; 201:216-224.
44. Wolfe S.A., Koppel D.A., Fogelfeld, L. and Grimes, S.R. Identification of protein binding to a conserved DNA sequence element within the proximal promoter of mammalian histone H1t genes. Mol. Biol. Cell 1992; 3 Supp:324A.
45. Kremer, E.J. and Kistler, W.S. Analysis of the promoter for the gene encoding the testis-specific histone H1t in a somatic cell line: Evidence for cell-cycle regulation and modulation by distant upstream sequences. Gene 1992; 110:167-173.
46. Wolfe, S.A. and Grimes, S.R. Protein-DNA interactions within the rat histone H4t promoter. J. Biol. Chem. 1991; 266:6637-6643.
47. Huh, N.E., Hwang, I.W., Lim, K., You, K.-H. and Chae, C.-B. Presence of a bidirectional S phase-specific transcription regulatory element in the promoter shared by testis-specific TH2A and TH2B histone genes. Nucl. Acids Res. 1991; 19:93-98.
48. Hwang, I., Kim, K. and Chae, C.B. S phase specific transcription regulatory elements are present in a replication independent testis-specific H2B histone gene. Mol.Cell.Biol. 1989; 9:1005-1013.
49. Hwang, I., Lim, K. and Chae, C.-B. Characterization of the S-phase specific transcription regulatory elements in a DNA replication-independent testis-specific H2B TH2B histone gene. Molec. Cell. Biol. 1990; 10:585-592.
50. Lim, K. and Chae, C.-B. Presence of a repressor protein for testis-specific H2B TH2B histone gene in early stages of spermatogenesis. J. Biol. Chem. 1992; 267:15271-15273.
51. Barberis, A., Superti-Furga, G. and Busslinger, M. Mutually exclusive interaction of the CCAAT-binding factor and of a displacement protein with overlapping sequences of a histone gene promoter. Cell 1987; 50:347-359.
52. Barberis, A., Superti-Furga, G. and Busslinger, M. The testis specific octamer binding protein of the sea urchin has a molecular weight of 85 kD. Nucl. Acids Res. 1989; 17:7114.
53. Hake, L.E. and Hecht, N.B. Utilization of an alternative transcription initiation site of somatic cytochrome c in the mouse produces a testis-specific cytochrome c mRNA. J. Biol. Chem. 1993; 268:4788-4797.
54. Evans, M.J. and Scarpulla, R.C. Interaction of nuclear factors with multiple sites in the somatic cytochrome c promoter. J. Biol. Chem. 1989; 264:14362-14368.
55. Langford, K.G., Zhou, Y., Russell, L.D., Wilcox, J.N. and Bernstein, K.E. Regulated expression of testis angiotensin-converting enzyme during spermatogenesis in mice. Biol. Reprod. 1993; 48:1210-1218.
56. Howard, T.E., Shai, S.-Y., Langford, K.G., Martin, B.M. and Berstein K.E. Transcription of testicular angiotensin-converting enzyme ACE is

initiated within the twelfth intron of the somatic ACE gene. Mol. Cell. Biol. 1990; 10:4294-4302.

57. Howard, T., Balogh, R., Overbeek, P. and Bernstein, K.E. Sperm-specific expression of angiotensin-converting enzyme ACE is mediated by a 91-base-pair promoter containing a CRE-like element. Molec. Cell. Biol. 1993; 13:18-27.
58. Oliva, R. and Dixon, G.H. Vertebrate protamine genes and the histone-to-protamine replacement reaction. Prog. Nucl. Acid Res. Molec. Biol. 1991; 40:25-94.
59. van de Hoorn, F.A. and Tarnasky, H.A. Factors involved in the regulation of the RT7 promoter in a male germ cell-derived in vitro transcription system. Proc. Natl. Acad. Sci. USA 1992; 89:703-707.
60. van Der Hoorn, F.A. Identification of the testis *c-mos* promoter: Specific activity in a seminiferous tubule-derived extract and binding of a testis-specific nuclear factor. Oncogene 1992; 7:1093-1099.
61. Mizuno, K. Goto, M., Masamune, Y. and Nakanishi, Y. A silencer-like *cis* element for the testis-specific phosphoglycerate-kinase-2-encoding gene. Gene 1992; 119:293-297.
62. Goto, M., Masamune, Y. and Nakanishi, Y. A factor stimulating transcription of the testis-specific *Pgk-2* gene recognizes a sequence similar to the binding site for a transcription inhibitor of the somatic-type *Pgk-1* gene. Nucl. Acids Res. 1993; 21:209-214.
63. Robinson, M.O., McCarrey, J.R. and Simon, M.I. Transcriptional regulatory regions of testis-specific PGK2 defined in transgenic mice. Proc. Natl. Acad. Sci. USA 1989; 86:8437-8441.
64. Gebara, M.M. and McCarrey, J.R. Protein-DNA interactions associated with the onset of testis-specific expression of the mammalian Pgk-2 gene. Molec. Cell Biol. 1992; 12:1422-1431.
65. Iannello, R.C., Kola, I. and Dahl, H.-H. M. Temporal and tissue-specific interactions involving novel transcription factors and the proximal promoter of the mouse Pdha-2 gene. J. Biol. Chem. 1993; 268:22581-22590.
66. Kurata, S., Matsumote, M., Hoshi, M. and Nakajima, H. Transcriptional control of the heme oxygenase gene during mouse spermatogenesis. Eur. J. Biochem. 1993; 217:633-638.
67. Galcheva-Gargova, Z., Taokeson, J.P., Karagyosov, L., Ebert, K.M. and Kilpatrick, D.L. The rat proenkephalin germ line promoter contains multiple binding sites for spermatogenic cell nuclear proteins. Molec. Endocrinol. 1993; 7:979-991.
68. Kilpatrick, D.L., Zinn, S.A., Fitzgerald, M., Higuchi, H., Savol, S.L. and Meyerhardt, J. Transcription of the rat and mouse proenkaphalin genes is intiated at distinct sites in spermatogenic and somatic cells. Mol. Cell. Biol. 1990; 10:3717-3726.
69. Kilpartick, D.L. and Millette, C.F. Expression of proenkephalin messenger RNA by mouse spermatogenic cells. Proc. Natl. Acad. Sci. USA 1986; 83:5015-5018.
70. Kew, D., Muffly, K.E. and Kilpatrick, D.L. Proenkephalin products are stored in the sperm acrosome and may function in fertilization. Proc. Natl. Acad. Sci. USA 1990; 87:1943-1947.
71. Borsook, D., Rosen, H. Collard, M., Dressler, H., Herrup, K., Comb, M.J. and Hyman, S.E. Expression and regulation of a proenkephalin

β-galactosidase fusion gene in the reproductive system of transgenic mice. Mol. Endocrinol. 1992; 6:1502-1512.
72. Donovan, D., Takemure, M., O'Hara, B., Brannock, M. and Uhi, G. Preproenkephalin promoter "cassette" confers brain expression and synaptic regulation in transgenic mice. Proc. Natl. Acad. Sci. USA 1992; 89:2345-2349.
73. Behringer, R.R., Crotty, D.A., Tennyson, V.M., Brinster, R.L., Palmiter, R.D. and Wolgemuth, D.J. Sequences 5' of the homeobox of the *Hox-1.4* gene direct tissue-specific expression of lacZ during mouse development. Development 1993; 117:823-833.
74. Wisniewski, J., Malezewski, M., Krawczyk, Z. and Gedamu, L. An upstream region of the rat spermatogenesis-specific heat-shock-like *Hst70* gene confers testis-specific expression in transgenic mice. Eur. J. Biochem. 1993; 212:137-143.
75. Nayernia, K., Burkhardt, E., Beimesche, S., Keime, S. and Engel, W. Germ cell-specific expression of a proacrosin-CAT fusion gene in transgenic mouse testis. Mol. Reprod. Devel. 1992; 31:241-248.
76. Michiels, F., Falkenburg, D., Muller, A., Hinz, U., Otto, U., Bellman, R., Glatzer, K.H., Brand, R., Bialojan, S. and Renkawitz-Pohl, R. Testis specific β tubulins are identical in *Drosophila melanogaster* and *D. hydei* but differ from the ubiquitous β1 tubulin. Chromosoma 1987; 95:387-395.
77. Trasler, J.M., Hake, L.E., Johnson, P.A., Alcivar, A.A., Millette, C.F. and Hecht, N.B. DNA methylation and demethylation events during meiotic prophase in the mouse testis. Mol. Cell. Biol. 1990; 10:1828-1834.
78. Alcivar, A.A., Trasler, J.M., Hake, L.E., Salehi-Ashtiani, K., Goldberg, E. and Hecht, N.B. DNA methylation and expression of genes coding for lactate dehydrogenase A and C during rodent spermatogenesis. Biol. Reprod. 1991; 44:527-535.
79. Lim, K. and Chae, C.-B. Presence of a repressor protein for testis-specific H2B (TH2B) histone gene in early stages of spermatogenesis. J. Biol. Chem. 1992; 267:15271-15273.
80. Ariel, M., McCarrey, J. and Cedar, H. Methylation patterns of testis-specific genes. Proc. Natl. Acad. Sci. USA 1991; 88: 2317-2321.
81. Salehi-Ashtiani, K., Widrow, R.J., Markert, C.L. and Goldberg, E. Testis-specific expression of a metallothionein I-driven transgene correlates with undermethylation of the locus in testicular DNA. Proc. Natl. Acad. Sci. USA 1993; 90:8886-8890.
82. Sapienza, C., Peterson, A.C., Rossant, J. and Balling, R. Degree of methylation of transgenes is dependent on gamete of origin. Nature 1987; 328:251-254.
83. Barlow, D.P. Methylation and imprinting: From host defense to gene regulation? Science 1993; 260:309-310.
84. DeChiara, T.M., Robertson, E.J. amd Efstratiadis, A. Parental imprinting of the mouse insulin-like growth factor II gene. Cell 1991; 64:849-859.
85. Barlow, D.P., Stoger, R., Hermann, B.G., Saito, K. and Schweifer, N. The mouse insulin-like growth factor type-2 receptor is imprinted and closely linked to the Tme locus. Nature 1991; 349:84-87.
86. Forejt, J. and Gregorova, S. Genetic analysis of genomic imprinting: An Imprintor-1 gene controls inactivation of the paternal copy of the mouse Tme locus. Cell 1992; 70:443-450.

87. Bartolomei, M.S., Zemel, S. and Tilghman, S.M. Parental imprinting of the mouse H19 gene. Nature 1991; 351:153-155.
88. Erickson, R.P. Post-meiotic gene expression. Trends Genet. 1990; 6:264-269.
89. Probst, F., Rosenberg, M.P. and van de Woude, G. Proto-oncogene expression in germ cell development. Trends Genetics 1988; 4:183-187.
90. Hake, L., Alcivar, A.A. and Hecht, N.B. Changes in mRNA length accompany translational regulation of the somatic and testis specific cytochrome c genes during spermatogenesis in the mouse. Devel. Biol. 1990; 110:249-257.
91. Garrett, J.F., Collard, M.W. and Douglass, J.O. Translational control of germ cell expressed mRNA imposed by alternative splicing: Opioid peptide gene expression in rat testis. Mol. Cell. Biol. 1989; 9:4381-4389.
92. Shaper, N.L., Wright, W.W. and Shaper, J.H. Murine β1,4-galactosyltransferase: Both the amounts and structure of the mRNA are regulated during spermatogenesis. Proc. Natl. Acad. Sci. 1990; 87:791-795.
93. Langford, K.G., Shai, S.-Y., Howard, T.E., Kovac, M.J., Overbeek, P.A. and Bernstein, K.E. Transgenic mice demonstrate a transgenic promoter for angiotensin-converting enzyme. J. Biol. Chem. 1991; 266:15559-15562.
94. Harduin-Lepers, A., Shaper, N.L., Mahoney, J.A. and Shaper, J.H. Murine β1,4-galactosyltransferase: round spermatid transcripts are characterized by an extended 5'-untranslated region. Glycobiology 1992; 2:361-368.
95. Fischman, K., Edman, J. Schakleford. G.M., Turner, J.A., Rutter, W.J. and Nir, U. A murine *fer* testis-specific transcript *ferT* encodes a truncated *fer* protein. Mol. Cell. Biol. 1990; 10:146-153.
96. Meijer, D., Hermans, A., von Lindern, M. van Agthoven, T., deKlein, A., Mackenbach, P., Grootegoed, A., Talarico, D., Della Valle, G. and Grosveld, G. Molecular characterization of the testis specific *c-abl* mRNA in mouse. EMBO J. 1987; 6:6041-6048.
97. Oyen, O., Myklebust, F., Scott, J.D., Cadd, J.D., McKnight, G.S., Hansson, V. and Jahnsen, T. Subunits of cyclic adenosine 3',5'-monophosphate-dependent protein kinase show differential and distinct expression patterns during germ cell differentiation: Alternative polyadenylation in germ cells gives rise to unique smaller-sized mRNA species. Biol. Reprod. 1990; 43:46-54.
98. Rao, S.M. and Howells, R.D. *Cis*-acting elements in the 5'-untranslated region of rat testis proenkephalin mRNA regulate translation of the precursor protein. J.Biol.Chem. 1993; 268: 22164-22169.
99. Kwon, Y.K. and Hecht, N.B. Binding of a phosphoprotein to the 3' untranslated region of the mouse protamine 2 mRNA temporally represses its translation. Molec. Cell. Biol. 1993; 13:6547-6557.
100. Kwon, Y.K. and Hecht, N.B. Cytoplamic protein binding to highly conserved sequences in the 3' untranslated region of mouse protamine 2 mRNA, a translationally regulated gene of male germ cells. Proc. Natl. Acad. Sci. USA 1991; 88:3584-3588.
101. Schluter, G., Schlicker, M. and Engel, W. A conserved 8 bp motif GCYATCAY in the 3' UTR of transition protein 2 as a putative target for a transcript stabilizing protein factor. Biochem. Biophys. Res. Comm. 1993; 197:110-115.
102. Kleene, K. PolyA shortening accompanies the activation of translation of

five mRNAs during spermiogenesis in the mouse. Development 1989; 106:367-373.

103. Gold, B., Fujimoto, H., Kramer, J.M., Erickson, R.P. and Hecht, N.B. Haploid accumulation and translational control of phosphoglycerate kinase-2 mRNA during mouse spermatogenesis. Devel. Biol. 1983; 98:392-399.

104. Alcivar, A.A., Hake, L.E., Kwon, Y.K. and Hecht, N.B. *junD* mRNA expression differs from *c-jun* and *junB* mRNA expression during male germinal cell differentiation. Mol. Reprod. Devel. 1991; 30:187-193.

105. Kleene, K.C. Multiple controls over the efficiency of translation of the mRNAs encoding transition proteins, protamines, and the mitochondrial capsule selenoprotein in late spermatids in mice. Devel. Biol. 1993; 159:720-731.

106. Kuhn, R., Schafer, U. and Schafer, M., Cis-acting regions sufficient for spermatocyte-specific transcription and spermatid-specific translational control of the *Drosophila melanogaster* gene *mst3gl-9.* EMBO J. 1988; 7:447-454.

107. Schafer, M.R., Kuhn, R., Bosse, R. and Schafer, U. A conserved element in the leader mediates post-meiotic translation as well as cytoplasmic polyadenylation of a *Drosophila* spermatocyte mRNA. EMBO J. 1990; 9:4519-4525.

108. Yanicostas, C., Vincent, A. and Lepasant, J.A. Transcriptional and post-transcriptional regulation contributes to the sex-regulated expression of two sequence related genes at the *janus* locus of *Drosophila melanogaster.* Mol. Cell Biol. 1989; 9:2526-2535.

109. Michiels, F., Buttgereit, D. and Renkawitz-Pohl, R. An 18-bp element in the 5' untranslated region of the *Drosophila* β2 tubulin mRNA regulates the mRNA level during postmeiotic stages of spermatogenesis. Eur. J. Cell Biol. 1993; 62:66-74.

110. Schafer, M., Borsch, D., Hulster, A. and Schafer, U. Expression of a gene duplication encoding conserved sperm tail proteins is translationally regulated in *Drosophila* melanogaster. Molec. Cell. Biol. 1993; 13:1708-1718.

111. Kempe, E., Muhs, B. and Schafer, M. Gene regulation in *Drosophila* spermatogenesis: Analysis of protein binding at the translational control element TCE. Devel. Genetics 1993; 14:449-459.

112. Lantz, V., Ambrosio, L. and Schedl, P. The *Drosophila orb* gene is predicted to encode sex-specific germ line RNA-binding proteins and has localized transcripts in ovaries and early embryos. Development 1992; 115:75-88.

CHAPTER 4

MOLECULES OF THE SOMATIC CELLS

INTRODUCTION

The major somatic cell types of the testis are the Sertoli cells of the seminiferous epithelium in intimate contact with the germ line (Fig. 4.1), the Leydig cells of the interstitium (Fig. 4.2), and the peritubular myoid cells which with the basement lamina form the outer wall of the seminiferous tubule (Fig. 4.3). Recently characterized molecules of these cells are discussed in this chapter, except for hormones and growth factors which are discussed in chapter 5. Potential interactions between these cells are considered in chapter 6.

SERTOLI CELLS

A large number of products of the Sertoli cell have been investigated, some cell-specific and some shared with other cell types. These products are often studied in whole-body irradiated rats lacking germ cells (Sertoli cell-enriched testes), in cultured purified Sertoli cells, or in *SXR* mutant mice devoid of germ cells. Well characterized products of the Sertoli cell include transport proteins, hormones, growth factors, enzymes, and molecules of the extracellular matrix.

TRANSPORT AND BINDING PROTEINS

One of the earliest recognized products which also serves as a useful Sertoli cell marker is *androgen binding protein (ABP)* reviewed in refs. 1 and 2. ABP is secreted by Sertoli cells, and binds testosterone or dihydrotestosterone with high affinity, carrying it to the epididymis.[1,3] ABP is an approximately 100 kDa dimer, composed of two subunits of 47 kDa or one of 47 kDa and one of 41 kDa.[1] The molecule shows heterogeneity due to oligosaccharide moieties.[4] ABP mRNA is maximal at Stages XIII to XI[5] coincident with maximal secretion of ABP protein.[6]

Transferrin is a glycoprotein of 70-80 kDa which transports iron and is made by many cell types. It is a major secretory product of Sertoli cells and transfers iron from the basal to the adluminal compartment.[7,8]

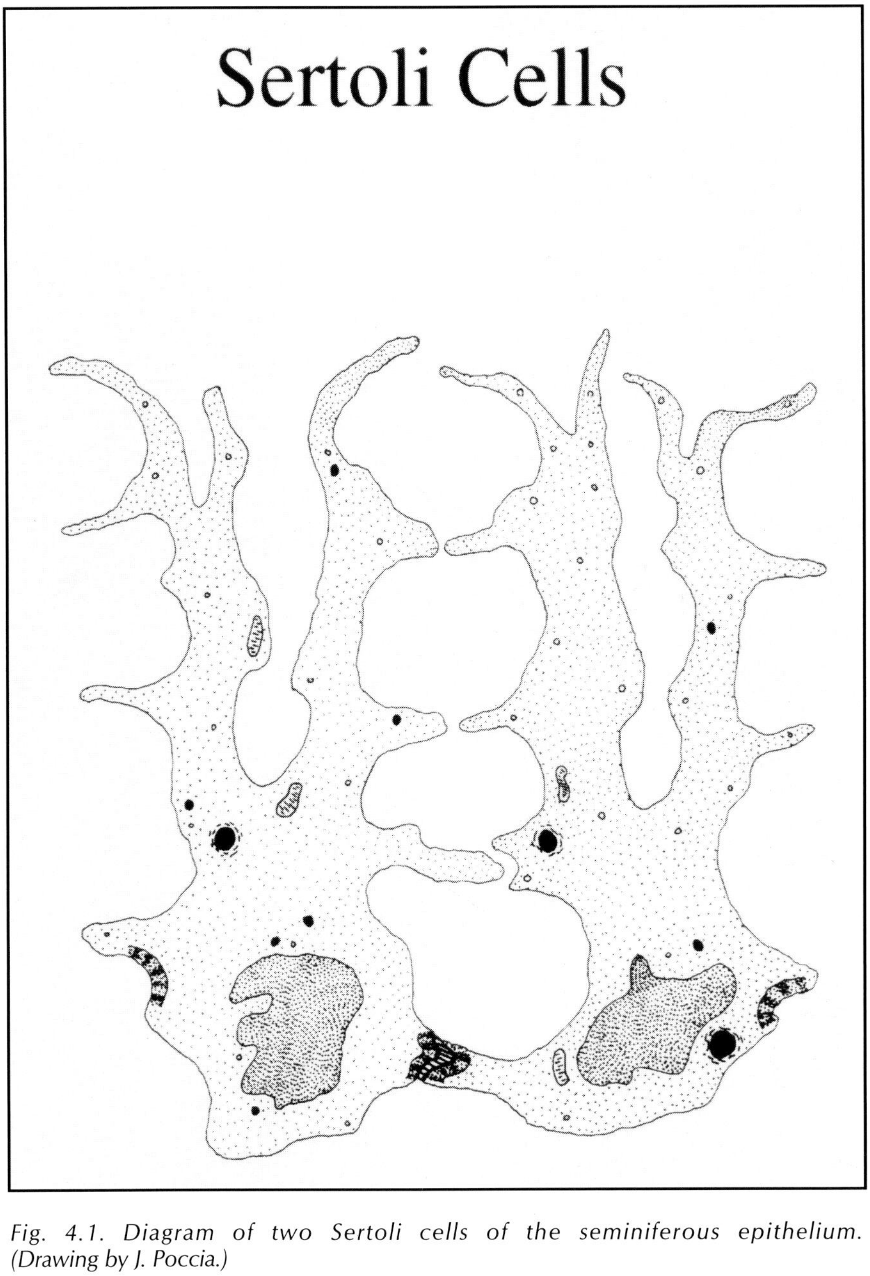

Fig. 4.1. Diagram of two Sertoli cells of the seminiferous epithelium. (Drawing by J. Poccia.)

Sertoli cell transferrin has the same amino acid sequence as liver transferrin, but a different glycosylation pattern.[1] Its N- and C-terminal halves are similar and it binds two irons and two carbonates.[9] The transferrin gene in human Sertoli cells has an inactive enhancer due to the binding of a nuclear protein.[10] An activating TATA-box binding factor has also been reported.[10] Transferrin mRNA expression is highest at Stages XII-XIV during meiotic division, but Sertoli cell *transferrin receptor* mRNA is maximal (5-7-fold over basal levels) at Stages VII to XI, at the time when preleptotene spermatocytes appear.[5] Sertoli transferrin receptors are

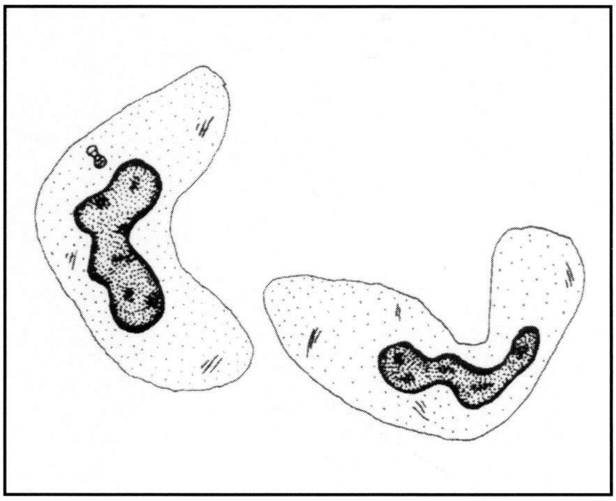

Fig. 4.2. Diagram of two Leydig cells of the interstitium. (Drawing by J. Poccia.)

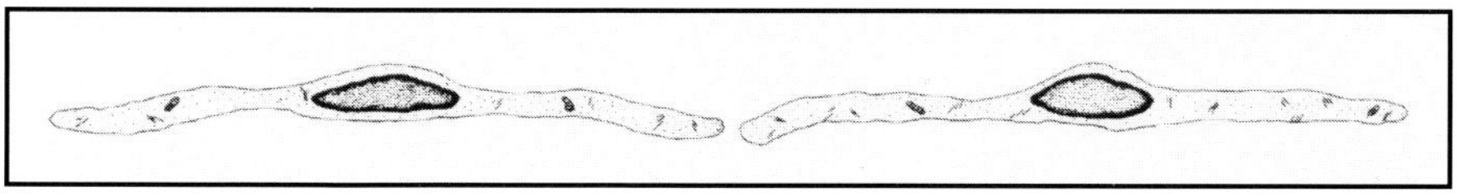

Fig. 4.3. Diagram of two peritubular myoid cells of the wall of the seminiferous epithelium. (Drawing by J. Poccia.)

located on the basal surface of the cells[11] and their induction may serve to promote iron accumulation when iron concentrations are low.[5] Both iron and transferrin are secreted into the adluminal space surrounding the germ cells. Transferrin receptors are found, progressively declining, on germ cells from spermatogonia to early spermatids, but not on late spermatids or spermatozoa.[1,12] A low molecular weight iron binding protein probably transfers iron to apotransferrin in the adluminal space which is then taken up by germ cells and stored as ferritin.[13] The transferrin receptor is a transmembrane glycoprotein, a dimer of 90 kDa subunits linked by disulfide bonds. Each subunit is predicted to contain a large extracellular portion, a single membrane spanning domain and a small cytoplasmic region.[14] Each subunit can bind one transferrin molecule. The human gene has 19 exons, is 31 kb long[14] and has TATA, CG-rich CRE and AP-1 sites in its promoter.[15,16] Sertoli cells in culture regulate transferrin receptor levels in response to iron.[17] Sertoli transferrin receptor mRNA has five regions which can serve as response elements to iron.[18]

Ceruloplasmin is a 140 kDa protein that transports Cu^{++}.[1,19] It may also promote the action of transferrin.[20]

SPARC (secreted protein acidic and rich in cysteine) is a 43 kDa protein produced by Sertoli cells related to osteonectin, BM-40 of the basement lamina and a mouse embryo endoderm protein.[21] It binds Ca^{++} and may be involved in Ca^{++} transport during spermatogenesis. It is also expressed in Leydig cells.[22,23]

Gamma-Glutamyl transpeptidase (γ-GTP) is a membrane-bound enzyme which catalyzes transfer of the γ-glutamyl group of glutathione to amino acids and peptides to aid in their transport across the plasma membrane.[24,25]

It is useful as a Sertoli cell marker since it is essentially absent from germ cells, peritubular cells, and Leydig and other interstitial cells.[2]

Hormones and Growth Factors

A number of hormones and growth factors are synthesized and secreted by Sertoli cells. These may be involved in autocrine or paracrine functions. They include anti-Mullerian hormone, activin, inhibin, follistatin, somatomedin C (IGF-I), interleukin-1, Leydig cell stimulatory factors, Leydig inhibitor factor, Sertoli cell secreted growth factor (SCSGF), EGF-like growth factor, transforming growth factors TGF-α and -β, and prohibitin. Sertoli cells express a number of receptors for hormones and growth factors including receptors for androgens, activin, follicle stimulating hormone FSH, nerve growth factor NGF, fibroblast growth factor FGF-β, insulin-like growth factor IGF-I, retinoic acid and *c-kit* ligand or Steel factor. These have been characterized to various degrees in the testis. Many are made by several other cell types. Those factors apparently specific to Sertoli cells are discussed below.

Anti-Mullerian hormone or Mullerian inhibiting substance (AMH or MIS) is a Sertoli secretory product responsible for regression in males of the Mullerian ducts (oviduct precursors). AMH is a member of the TGF-β family.[26] It is expressed early in male fetal development (e.g., in immature Sertoli cells at 12.5 d post-coitum in mice). Germ cells are not required for its expression. AMH levels decline when Sertoli cells mature, but before the blood-testis barrier is formed. AMH is a multimeric glycoprotein of 145 kDa to 235 kDa which reduces to 72 kDa subunits.[1] It is synthesized as a precursor of 575, 560, 553 and 555 amino acids in bovine, human, rat and mouse Sertoli cells respectively. The four mammalian proteins share about 75% amino acid sequence homology. Their C-terminal regions are 95% homologous, similar to TGF-β, and biologically active. Two to three potential N-linked glycosylation sites are found but are not essential for activity. The protein has an N-terminal signal sequence. The mRNAs have short 5' and 3' untranslated regions suggesting little if any translational regulation. The mRNAs are undetected in germ cells or other somatic testicular cells. For a review of AMH, see ref 26.

Human and bovine AMHs are secreted as 140 kDa homodimers. Although the biological function of AMH in regression of Mullerian ducts is well documented, other roles have been suggested. Because of its continued expression in males after duct regression, it has been suggested to have a role in germ cell maturation.[26] AMH may cause mitotic arrest of spermatogonia which enter meiosis when AMH levels decline.[26] Addition of anti-AMH antibody to cultured neonatal mouse testis inhibits AMH-dependent development from gonocytes to Type-A spermatogonia.[27] A receptor for AMH has not yet been cloned or characterized. AMH may be regulated by testis determining factor (TDY, see chapter 6) because it is the first known Sertoli cell product.

A human mutation called persistent Mullerian duct syndrome (PMDS) is due to defective AMH in males. One individual had codon 358 changed to a stop codon resulting in a truncated protein which is unstable.[28] Another mutation results from a single substitution of Trp for Arg at codon 377.[29]

Sertoli cell secreted growth factor (SCSGF) is one of two growth factors apparently specific to the Sertoli cell.[30] Activity resides in a 2 kDa and an 8 kDa fraction purified from Sertoli cell conditioned medium.[31] SCSGF stimulates the proliferation of several cell types including a cell-line (A431) which is not affected by TGFs, FGF, PDGF, IL-1b or IGF-like[32] or other Sertoli cell secretions and which thus serves as a convenient assay system for SCSGF.[30] SCSGF competes for EGF receptor binding.[31]

Interleukin-1α is the second growth factor made exclusively by Sertoli cells in the testis.[30] It is a 17 kDa protein in rat testis and is absent from peritubular conditioned medium.[33,34] IL-1α stimulates DNA synthesis in spermatogonia and meiotic spermatocytes.[35-38] IL receptors are found on interstitial cells[39] suggesting a role in Leydig cell regulation.

Extracellular Matrix, Junctional Proteins, and Extracellular Proteases

The seminiferous epithelium possesses an intricate and dynamic structure dependent on Sertoli cells. Sertoli cells attach to basement lamina, to each other, and to developing germ cells. Interactions with germ cells must be modulated during spermatogenesis to allow proliferation of spermatogonia and to release germ cells which move towards the lumen. Extensive tissue remodeling occurs during spermatogenesis involving proteins of the extracellular matrix and junctional complexes.

Sertoli cells synthesize *Type IV collagen* and *laminin*, the main components of their basement lamina.[40] These molecules are presumably secreted from the basal compartment of the germinal layer.[1] Sertoli cells make *Type IV collagenase* and other metalloproteins as well.[41] Sertoli cells also make an inhibitor of metalloproteinase Type-2, *TIMP-2*, which can inhibit Type IV collagenase.[42] Antibodies to laminin and entactin, but not antibodies to Types I or IV collagen, fibronectin or heparan sulfate proteoglycan, inhibit seminiferous cord formation in vitro.[40] Laminin/entactin-mediated promotion of cord formation is inhibited by RGD (arginine-glycine-aspartic acid) peptides characteristic of certain extracellular matrix proteins recognized by integrins, but Sertoli cell attachment to extracellular matrix in vitro is not.[43] The pentapeptide YIGSR (tyrosine-isoleucine-glycine-serine-arginine), corresponding to a cell binding domain of laminin, inhibits cord formation and Sertoli attachment.

Ca^{++}-dependent cadherins, Ca^{++}-independent cell-adhesion molecules (CAMs) and integrins are all found in the seminiferous epithelium.[44] A 79 kDa protein associated with Sertoli cells and germ cells and three other proteins found on Sertoli cell surfaces may aid in binding Sertoli cells and pachytene spermatocytes.[45] Several *laminin binding proteins* have been isolated from Sertoli cell surfaces.[40,46] Many localize to the basolateral plasma membrane. At least two are related to integrins. Two *cadherin* mRNAs have been localized to the testis, but their cellular origin remains unspecified.[47] N-Cad and P-Cad mRNAs were detected in rat testis but not E-Cad mRNAs. The temporal pattern of mRNA expression suggests that P-Cad might be involved in testicular development and N-Cad in later spermatogenesis. Antibodies raised against N-cadherin have been used to immunolocalize the molecule to the surfaces of Sertoli cells and spermatogenic cells and detect it on Western blots from plasma membranes of the cells.[48] The antibody reduced cell adhesion between Sertoli and germ line cells in an in vitro binding assay.

Rat Sertoli cells make proteoglycans containing chondroitin, heparan, or mixed glycosaminoglycans, and peritubular cells make proteoglycans with chondroitin.[49] *Testicans* are proteoglycan precursors to human seminal plasma peptides linked to heparan/chondroitin-sulfate glycosaminoglycans.[50] The testican gene was cloned from a human testicular library. The sequence predicts a 439-amino acid core protein precursor. It has an N-terminal hydrophobic domain with a signal sequence and four regions similar to the osteonectin precursor involved in cell-cell and cell-matrix interactions, One of these contains a Kazal-like domain found in rat agrin, a basal lamina component. The molecule also contains a 46-amino acid stretch rich in cysteine similar to the Type-I thyroglobulin repeats involved in cell-cell or cell-matrix interactions and similar to insulin-like growth factor binding domains. Therefore the molecule has potential properties of cell, matrix and growth factor binding.

Testins I and *II* [51] are almost identical glycoproteins secreted by Sertoli cells which are associated with Sertoli-Sertoli cell junctions and Sertoli cell-spermatid junctions.[52] Testins are synthesized and secreted by Sertoli cells in vitro.[52-54] Formerly called *CMB22* and *CMB23,* their release is simulated by testosterone and inhibited by germ cells.[51]

Several proteases and protease inhibitors are associated with Sertoli cells which may be involved in modulating attachments between cells during tissue remodeling. *Plasminogen activator* is a protease which converts plasminogen to plasmin. It is maximally secreted at Stages VII and VIII when spermatocytes pass through the tight junctions and spermatids are released. Therefore it has been postulated to restructure the Sertoli-germ cell layer.[55,56]

Cathepsin L is a lysosomal cysteine protease involved in intracellular protein degradation.[57,58] It has a mannose-6-phosphate recognition signal which would serve to target it to the lysosomal compartment.[59,60] As isolated from Sertoli cell conditioned medium, it is a 32 kDa protein of two subunits: 27 kDa and 5 kDa. The 27 kDa chain is identical to the heavy chain of rat kidney cathepsin L. Originally identified as *cyclic protein 2,* a protein secreted in a cyclical fashion, it is maximally secreted at Stage VI.[1,59,60] A partial clone of the gene corresponds to procathepsin L.[61] The major 1.7 kb mRNA is localized to Sertoli cells in rat and peaks at Stages IV to VII.[62]

Cystatin C is a 200-amino acid cysteine protease inhibitor which can inhibit cathepsin L and is secreted in glycosylated and unglycosylated forms from rat Sertoli cells in vitro.[63] Its concentration is inversely correlated with cathepsin L.[64] It is also secreted by Leydig cells. It has been suggested to protect testis cells from secreted lysosomal enzymes released during degeneration of germ cells or to function in tissue remodeling. Sertoli cells have a 700 nucleotide transcript for cystatin C.[65] Although no protein was detected in germ cells, a 500 nt transcript was present in a population of spermatids and spermatocytes.[65]

Alpha-2-Macroglobulin is a non-specific protease inhibitor which may also be involved in tissue remodeling[66,57] It is actively synthesized and secreted by Sertoli cells. It binds inhibin and activin, cytokines and growth factors.[67]

LRP, liver-regulating protein-like molecule, is an integral plasma membrane protein found in regions of spermatocyte-Sertoli cell adherence. Antibodies to LRP block pachytene spermatocyte-stimulated transferrin production by Sertoli cells.[68,69]

Other Sertoli Proteins

Sulfated glycoproteins-1 and *-2* are the major secretory products of Sertoli cells. Their functions remain unknown. SGP-1 is a 140 kDa glycoprotein made of two identical subunits. SGP-2 is made as a 64 kDa glycosylated precursor with a 21-amino acid signal sequence.[70] It is cleaved to a 47 kDa α-subunit with four glycosylation sites and a 34 kDa β-subunit with two glycosylation sites. The subunits are disulfide linked. Both SGPs are N-glycosylated and very acidic, containing sulfated sugar residues and sialic acid.

SGP-1, also known as *proaposin*, may be a carrier for lipids.[4,71,72] It binds sphingolipids and is incorporated into the sperm plasma membrane.

SGP-2 is the same molecule as several independently investigated proteins including the protein coded by *testosterone-repressed prostate message-2 (TRPM-2)* from rat prostate, *clusterin* and *S35-S45 heterodimeric protein*.[73] It may also function to carry lipids and can be found on the acrosome and tail of sperm.[71] The message is constitutively expressed in adult Sertoli cells, appearing in 7-14 day old rats.[73] An antigenically related protein is also expressed in meiotic stages in the germ line.[74] SGP-2 mRNA is 1.5 kb in developing testis and 1.8 kb in the adult, probably due to differential polyadenylation rather than differential splicing. In synchronized testes, the highest mRNA levels are found at Stage VIII where the message is associated with polysomes.[75] Secretion is not stage-regulated in Sertoli cell cultures.[75]

Rat clusterin has been reported to be a heterogeneous glycoprotein of 85 kDa made of a 43 kDa α- and a 35 kDa β-subunit present in almost all tissues.[57] It exists in multiple molecular forms which differ between testis, epididymis, prostate and seminal vesicle.[76] It induces aggregation of sperm (hence its name),[77] probably because of its high carbohydrate content.[1] Clusterin antigens are detected at the acrosome and outer dense fiber region of the sperm tail. Sequence homology with apolipoprotein suggests it may be involved in membrane synthesis.[70] SGP-2 is found in seminiferous tubule fluid and on membranes of late spermatids.[71]

Testibumin is a 68 kDa protein of unknown function related to albumin.[78] It is useful as a Sertoli cell marker.[57] *CMB-2* is a 22 kDa protein of unknown function whose secretion is stimulated by FSH but not testosterone.[79]

Cellubrevin, a member of the synaptobrevin family, is expressed at high levels in Sertoli cells.[80] Synaptobrevin-2 is a protein of small synaptic vesicles of neurons cleaved by metalloproteases such as tetanus toxin light chain. Transgenic mice expressing the tetanus toxin light chain show severely altered spermatogenesis, few mature sperm cells, increased numbers of Leydig cells and abnormal distributions of actin filaments in their Sertoli cells.[80]

Glutathione-S-transferase (GST) consists of four isozymes expressed in Leydig and Sertoli cells. GST is involved in metabolism of carcinogens and cytotoxic compounds and may protect germ cells during spermatogenesis. An isozyme, GST Pi, is not expressed in normal spermatogenesis, but is in teratomas, seminomas and neoplastic germ cells.[81]

Subunits of *G proteins* Gsα and Gβ and low levels of Gi have been immunolocalized in Sertoli cells (as well as Leydig and peritubular cells).[82]

The Wilms' tumor gene, *WT1,* codes for a transcription factor[83] expressed in somatic tissue of the gonadal ridge[84] and in Sertoli cells. Its testis-specific transcript is 2.1 kb whereas the normal somatic transcript is 3.1 kb.[85] It may play a role in preventing proliferation of adult Sertoli cells.

LEYDIG CELLS

Leydig cells are the most studied of the interstitial cells. They are found increasingly in small clusters during differentiation.[86] Follicle-stimulating hormone and luteinizing hormone promote Leydig cell differentiation. FSH probably acts indirectly through Sertoli cells.

The most important product of the Leydig cells is *testosterone* which is primarily controlled by LH, both in fetal and adult cells. Leydig precursor cells have androgen receptors. Dihydrotestosterone stimulates testosterone production in Leydig cell precursors. Steroid hormones and their receptors are discussed in chapter 5.

IGF-I, inhibin and TGF-α all stimulate steroidogenesis in Leydig cells, but all are also made by Leydig cells.[87,88] *Prohibitin,* a negative regulator of cell proliferation, is constitutively expressed by adult Leydig cells.[89] Leydig cells also make *β-endorphins* and *corticotrophin releasing factor.*

Apolipoprotein E is made by Leydig cells and a variety of other cell types.[90] It is an arginine-rich glycosylated protein of 299 amino acids. Most testicular expression of Apo E is in interstitial cells with diffuse expression throughout the seminiferous tubules. It is the protein component of very-low density lipoproteins (VLDL) and chylomicrons. The ligand for LDL receptor mediates liver uptake of cholesterol. Apo E may be involved in lipid distribution including delivery of cholesterol to other cells of the testis or in control of levels of steroids in the steroidogenic Leydig cell. A Leydig cell line of mouse expresses ApoE and should prove increasingly useful.

Leydig cells are also a source of *insulin-like growth factor-I* (IGF-I) as are other interstitial cells.[91] Leydig cells and interstitial macrophages produce *proopiomelanocortin.*[92] Leydig cells have been reported to produce *platelet derived growth factor* (PDGF) and *PDGF receptor.*[93] A member of the insulin-like peptide family, *Ley I-L,* is expressed exclusively in Leydig cells.[94] It is made as a proprotein of 131 amino acids with a 24-amino acid signal peptide and is cleaved into a 32-amino acid B peptide and a 26-amino acid A peptide.

PERITUBULAR CELLS

Peritubular cells contribute a major part of the outer wall of the seminiferous tubules. In addition to the Sertoli cells, they may contribute to the blood-testis barrier. A basement lamina separates the peritubular cells from the Sertoli cells.

Androgens stimulate peritubular cells to make *PModS*,[95] a factor which exhibits the most powerful effects on Sertoli cells of any natural compound to date, even more than FSH.[96] A 56 kDa PModS (A) is processed from a 59 kDa PModS (B).[97] Both are bioactive. PModS inhibits induction of aromatization activity in Sertoli cells in culture[98] and stimulates secretion of ABP.[95,99]

PModS appears to act through transcription factors to stimulate Sertoli transferrin expression.[95] It has no effect on transferrin mRNA stability and requires protein synthesis. Treating Sertoli cells with antisense c-*fos* oligonucleotides inhibits the effects of PModS on transferrin expression. PModS results in an increase of Sertoli cell c-*fos* but not c-*jun*.

Peritubular cells secrete *Type I* and *Type IV collagens, fibronectin,* basement membranes,[1] and proteoglycans containing chondroitin sulfate.[49] They also secrete several Ca^{++}-dependent metalloproteases in culture including *Type IV collagenase.*[100] Peritubular cells secrete a *plasminogen activator inhibitor (PAI-1)*[101] which inhibits plasminogen activity originating from Sertoli cells in coculture.[102] Peritubular cells also secrete *TGF-α,*[103] *TGF-β*[104] and *IGF-I,*[105] molecules discussed in chapter 5.

SUMMARY

The Sertoli cells are the best characterized somatic cells of the testis. They are secretory and phagocytic. They have many functions and are the chief mediators of the germ cell environment. They make a variety of growth factors, hormones, and transport proteins and serve to create and remodel the extracellular matrix. Leydig cells make testosterone and a number of other factors critical for spermatogenesis. Peritubular cells contribute to the blood-testis barrier and make a powerful substance which alters Sertoli cell activity. Interactions of the somatic cells of the testis are discussed in chapter 6.

REFERENCES

1. Griswold, M.D. Protein secretion of Sertoli cells. Int. Rev. Cytol. 1988; 110:133-156.
2. Tindall, D.J., Rowley, D.R., Murthy, L., Lipshultz, L.I. and Chang, C.H. Structure and biochemistry of the Sertoli cell. Int. Rev. Cytol. 1985; 94:127-149.
3. Tindall, D.J., Miller, D.A. and Means, A.R. Characterization of androgen receptors in Sertoli cell enriched testis. Endocrinol. 1977; 101:13-23.
4. Collard, M.W., Sylvester, S.R., Tsuruta, J.K. and Griswold, M.D. Biosynthesis and molecular cloning of sulfated glycoprotein 1 secreted by rat Sertoli cells: Sequence similarity with the 70-kDa precursor to sulfatide/GM1 activator. Biochemistry 1988; 27:4557-4564.
5. Linder, C.C., Heckert, L.L., Roberts, K.P., Kim, K.H. and Griswold, M.D. Expression of receptors during the cycle of the seminiferous epithelium.

Ann. N.Y. Acad. Sci. 1991; 637:313-321.

6. Parvinen, M. Regulation of the seminiferous epithelium. Endocrinol. Rev. 1982; 3:404-417.
7. Skinner, M.K. and Griswold, M.D. Sertoli cells synthesize and secrete a transferrin-like protein. J. Biol. Chem. 1980; 255:9523-9525.
8. Monet-Kuntz, C., Guillou, F., Fontaine, I. and Combarnous, Y. Purification of ovine transferrin and study of its secretion in enriched cultures of ovine Sertoli cells. J. Reprod. Fertil. 1992; 94:189-201.
9. Sylvester, S.R. and Griswold, M.D. Molecular biology of iron transport in the testis. In: deKretser, D. ed. Molecular Biology of the Male Reproductive System. NY: Acad. Press, 1993:311-327.
10. Guillou, F., Zakin, M.M., Part, D., Bossier, F. and Schaffer, E. Sertoli cell-specific expression of the human transferrin gene. J. Biol. Chem. 1991; 266:9876-9884.
11. Morales, C. and Clermont, Y. Receptor mediated endocytosis of transferrin by Sertoli cells of the rat. Biol. Reprod. 1986; 35:393-405.
12. Segretain, D., Egloff, M., Gerard, N., Pineau, C. and Jegou, B. Receptor-mediated and absorptive endocytosis by male germ cells of different mammalian species. Cell Tiss. Res. 1992; 268:471-478.
13. Wauben-Penris, P.J.J., Veldscholte, J., van der Ende, A. and van der Donk, H.A. The release of iron by Sertoli cells in culture. Biol. Reprod. 1988; 38:1105-1113.
14. Schneider, C. Owen, M.J., Banville, D. and Williams, J.G. Primary structure of human transferrin receptor deduced from the mRNA sequence. Nature 1984; 311:675-678.
15. Beard, P., Offord, E., Paduwat, N. and Bruggmann, H. SV40 activates transcription from the transferrin receptor promoter by inducing a factor which binds to the CRE/AP-1 recognition sequence. Nucl. Acids Res. 1991; 19:7117-7123.
16. Casey, J.L., Di Jeso, B., Rao, K.K., Rouault, T.A., Klausner, R.D. and Hartford, J.B. Deletional analysis of the promoter region of the human transferrin receptor gene. Nucl. Acids Res. 1988; 16:629-646.
17. Roberts, K.P. and Griswold, M.D. Characterization of rat transferrin receptor cDNA: The regulation of transferrin receptor mRNA in testes and in Sertoli cells in culture. Endocrinol. 1990; 44:531-542.
18. Casey, J.L., Koller, D.M., Ramin, V.C., Klausner, R.D. and Harford, J.B. Iron regulation of transferrin receptor mRNA levels requires iron-responsive elements and a rapid turnover determinant in the 3' untranslated region of the mRNA. EMBO J. 1989; 8:3693-3699.
19. Wright, W.W., Musto, N.A., Mather, J.P. and Bardin, C.W. Sertoli cells secrete both testis-specific and germ proteins. Proc. Natl. Acad. Sci. USA 1981; 78:7565-7569.
20. Stallard, B.J. and Griswold, M.D. Germ cell regulation of Sertoli cell transferrin mRNA levels. Mol. Endocrinol. 1990; 4:393-401.
21. Cheng, C.Y. Purification of a calcium binding protein rat SPARC from primary Sertoli cell-enriched culture medium. Biochem. Biophys. Res. Comm. 1990; 167:1393-1399.
22. Howe, C.C., Overton, G.C., Sawicki, J., Solter, D., Stein, P. and Strickland, S. Expression of SPARC-osteonectin transcript in murine embryos and gonads. Differentiation 1988; 37:20-25.

23. Vernon, R.A. and Sage, H. The calcium-binding protein SPARC is secreted by Leydig and Sertoli cells of the adult mouse testis. Biol. Reprod. 1989; 40:1329-1340.
24. Schteingart, H.F., Rivarola, M.A. and Cigorraga, S.B. Hormonal and paracrine regulation of γ-glutamyl transpeptidase in rat Sertoli cells. Mol. Cell. Endocrinol. 1989; 67:73-80.
25. Risley, M.S. and Morse-Gaudio, M. Comparative aspects of spermatogenic cell metabolism and Sertoli cell function in *Xenopus* and mammals. J. Exp. Zool. 1992; 261:185-193.
26. Cate, R.L. and Wilson, C.A. Mullerian-inhibiting substance. In: Gwatkin, R.G.B. ed. Genes in Mammalian Reproduction. New York: Wiley-Liss, 1993:185-205.
27. Zhou, B., Watts, L.M. and Hutson, J.M. Germ cell development in neonatal mouse testes in vitro requires Mullerian inhibiting substance. J. Urol. 1993; 150:613-616.
28. Guerrier, D., Tran, D., van der Winden et al. The persistent Mullerian duct syndrome: A molecular approach. J. Clin. Endocrinol. Metab. 1989; 68:46-52.
29. Josso, N. Anti-Mullerian hormone and Sertoli cell function. Horm. Res. 1992; 38 Supp 2:72-76.
30. Shubhada, S., Glinz, M. and Lamb, D.J. Sertoli cell secreted growth factor. Cellular origin, paracrine and endocrine regulation of secretion. J. Androl. 1993; 14:99-109.
31. Holmes, S.D., Spotts, G. and Smith, R.G. Rat Sertoli cells secrete a growth factor that blocks epidermal growth factor EGF binding to its receptor. J. Biol. Chem. 1986; 261:4076-4080.
32. Leung, T., How, B.E., Manser, E. and Lim, L. Germ cell β-chimaerin, a new GTPase-activating protein for p21rac, is specifically expressed during the acrosomal assembly stage in rat testis. J. Biol. Chem. 1993; 268:3813-3816.
33. Syed, V., Soder, O., Arver, S., Lindh, M., Khan, S. and Ritzen, E.M. Ontogeny and cellular origin of an interleukin-1-like factor in the reproductive tract of the male rat. Int. J. Androl. 1988; 11:437-447.
34. Sun, X.-R., Hedger, M.P. and Risbridger, G.P. The effect of testicular macrophages and interleukin-1 on testosterone production by purified adult rat Leydig cells cultured under in vitro maintenance conditions. Endocrinology 1993; 132:186-192.
35. Gerard, N., Syed, V., Bardin, C.W., Genetet, N. and Jegou, B. Sertoli cells are the site of interleukin-α1 synthesis in rat testis. Mol. Cell Endocrinol. 1991; 82:R13-R16.
36. Pollanen, P., Soder, O. and Parvinen, M. Interleukin-1 α stimulation of spermatogonial proliferation in vivo. Reprod. Fertil. 1989; 1:85-87.
37. Soder, O., Syed, V., Callard, J., Toppari, J., Pollanen, P., Parvinen, M., Froysa, B. and Ritzen, E.M. Production and secretion of an interleukin-1-like factor. Int. J. Androl. 1991; 14:223-231.
38. Parvinen, M., Soder, O., Mali, P., Froysa, B. and Ritzen, E.M. In vitro stimulation of stage-specific DNA synthesis in rat seminiferous tubule segments by interleukin-1α. Endocrinol. 1991; 129:1614-1620.
39. Takao, T., Mitchell, W.M., Tracey, D.E. and DeSouza, E.B. Identification of interleukin-1 receptors in mouse testis. Endocrinol. 1990; 127:251-258.

40. Dym, M. Basement membrane regulation of Sertoli cell function. In: Baccetti, B, ed. Comparative Spermatology 20 Years After. NY: Raven, 1991:632-642.
41. Sang, Q.X., Stretler-Stevenson, W.G., Liotta, L.A. and Byers, S.W. Identification of Type IV collagenase in rat testicular cell culture: Influence of peritubular-Sertoli cell interactions. Biol. Reprod. 1990; 43:956-9664.
42. Ailenberg, M., Stetler-Stevenson, W.G. and Fritz, I.B. Secretion of latent type IV procollagenase and active type IV collagenase by testicular cells in culture. Biochem. J. 1991; 279:75-80.
43. Hadley, M.A., Weeks, B.S., Kleinman, H.K. and Dym, M. Laminin promotes formation of cord-like structures by Sertoli cells in vitro. Devel. Biol. 1990; 140:318-327.
44. Pelletier, R.M. and Byers, S.W. The blood testis barrier and Sertoli cell junctions: Structural considerations. Micros. Res. Tech. 1992; 20:3-33.
45. Newton, S.C. and Millette, C.F. Sertoli cell plasma membrane polypeptides involved in spermatogenic cell-Sertoli cell adhesion. J. Androl. 1992; 13:160-171.
46. Davis, C.M., Papadoupoulos, V., Jia, M., Yamada, Y., Kleinman, H.K. and Dym, M. Identification and partial characterization of laminin binding proteins in immature rat Sertoli cells. Exp. Cell Res. 1991; 193:262-273.
47. Cyr, D.G., Blaschuk, O.W. and Robaire, B. Identification and developmental regulation of cadherin messenger ribonucleic acids in the rat testis. Endocrinol. 1992; 131:139-145.
48. Newton, S.C., Blaschuk, O.W. and Millette, C.F. N-cadherin mediates Sertoli cell-spermatogenic cell adhesion. Devel. Dynamics 1993; 197:1-13.
49. Skinner, M.K. and Fritz, I.B. Testicular peritubular cells secrete a protein under androgen control that modulates Sertoli cell functions. Proc. Natl. Acad. Sci. USA 1985; 82:114-118.
50. Alliel, P.M., Perin, J.-P., Jolles, P. and Bonnet, F.J. Testican, a multidomain testicular proteoglycan resembling modulators of cell social behaviour. Eur. J. Biocehm. 1993; 214:347-350.
51. Jegou, B., Pineau, C. Velez de la Calle, J.F., Touzalin, A.-M., Bardin, C.W. and Cheng, C.Y. Germ cell control of testin production is inverse to that of other Sertoli cell products. Endocrinol. 1993; 132:2557-2562.
52. Zong, S.D., Bardin, C.W., Phillips, D. and Cheng, C.Y. Testins are localized to the junctional complexes of rat Sertoli cell and epididymal cells. Biol. Reprod. 1992; 47:568-572.
53. Cheng, C.Y., Grima, J., Stahler, M.S., Lockshin, R.A. and Bardin, C.W. Testins are structurally related Sertoli cell products whose secretion is tighty coupled to the presence of germ cells. J. Biol. Chem. 1989; 264:21386-21393.
54. Grima, J. Pineau, C., Bardin, C.W. and Cheng, C.Y. Rat Sertoli cell clusterin, α_2-macroglobulin, and testins: biosynthesis and differential regulation by germ cells. Mol. Cell. Endocrinol. 1992; 89:127-140.
55. LaCroix, M., Smith, F.E. and Fritz, I.B. Secretion of plasminogen activator by Seroli-cell enriched cultures. Mol. Cell. Endocrinol. 1977; 9:227-236.
56. LaCroix, M., Parvinen, M. and Fritz, I.B. Localization of testicular plasminogen activator in discrete portions of Stages VII and VIII of the

seminferous tubule. Biol.Reprod. 1981; 25:143-146.

57. Cheng, C.Y., Zwain, I., Grima, J. and Bardin, C.W. Structure and functions of Sertoli cell proteins. In: Baccetti, B. ed. Comparative Spermatology 20 Years After. NY: Raven, 1991:617-624.
58. Ishidoh, K., Towatari, T., Imajoh, S., Kawasaki, H., Kominami, E., Katunuma, N. and Suzuki, K. Molecular cloning and sequencing of complementary DNA for rat cathepsin L. FEBS Lett. 1987; 223:69-73.
59. Wright, W.W., Parvinen, M., Musto, N.A., Gunsalus, G.L., Phillips, D.M., Mather, J.P. and Bardin, C.W. Identification of stage specific proteins synthesized by rat seminiferous tubules. Biol. Reprod. 1983; 29:257-270.
60. Shabanowitz, R.B., DePhilip, R.M., Crowell, J.A., Tres, L.L. and Kierszenbaum, A.L. Temporal appearance and cyclic behavior of Sertoli-cell specific secretory protein during the development of rat seminiferous tubules. Biol. Reprod. 1986; 35:745-760.
61. Erickson-Lawrence, M., Zabludoff, S.D. and Wright, W.W. Cyclic protein-2, a secretory product of rat Sertoli cells, is the proenzyme form of cathepsin L. Mol. Endocrinol. 1991; 5:1789-1798.
62. Maguire, S.M., Millar, M.R., Sharpe, R.M. and Saunders, P.T.K. Stage-dependent expression of mRNA for cyclic protein 2 during spermatogenesis is modulated by elongate spermatids. Molec. Cell. Endocrinol. 1993; 94:79-88.
63. Esnard, A., Esnard, F., Guillow, F. and Gauthier, F. Production of the cysteine proteinase inhibitor cystatin C by rat Sertoli cells. FEBS Lett. 1992; 300:131-135.
64. Linder, C.C., Heckert, L.L., Roberst, K.P., Kim, K.H. and Griswold, M.D. Expression of receptors during the cycle of the seminiferous epithelium. Ann. N.Y. Acad. Sci. 1991; 637:313-321.
65. Tsuruta, J.K., O'Brien, D.A. and Griswold, M.D. Sertoli cell and germ cell cystatin C: Stage-dependent expression of two distinct messenger ribonucleic acid transcripts in rat testes. Biol. Reprod. 1993; 49:1045-1054.
66. Cheng, C.Y., Grima, J., Stahler, M.S., Guglielmotti, A., Silvestrini, B. and Bardin, C.W. Sertoli cell synthesizes and secretes a protease inhibitor, α_2 macroglobulin. Biochemistry 1990; 29:1063-1068.
67. Cheng, C.Y., Grima, J., Pinaeau, C., Stahler, M.S., Jegou, B. and Bardin, C.W. α_2-Macroglobulin protects the seminiferous epithelium during stress. In: Sheppard, K.E. Boublik, J.H. and Funder, J.W. eds. Stress and Reproduction. NY: Raven Press, 1992:123-133.
68. Jegou, B. The Sertoli-germ cell communication network in mammals. Int. Rev. Cytol. 1993; 147:25-96.
69. Jegou, B. and Sharpe, R.M. Paracrine mechanisms in testicular control. In: deKretser, D. ed. In, Molecular Biology of the Male Reproductive System. NY: Acad. Press, 1993:271-310.
70. Collard, M.W. and Griswold, M.D. Biosynthesis and molecular cloning of sulfated glycoprotein 2 secreted by rat Sertoli cells. Biochem. 1987; 26:3297-3303.
71. Sylvester, S.R., Skinner, M.K. and Griswold, M.D. A sulfated glycoprotein synthesized by Sertoli cells and by epididymal cells is a component of the sperm membrane. Biol. Reprod. 1984; 31: 1087-1101.
72. Skinner, M.K. Cell-cell interactions in the testis. Endocrinol. Rev. 1991;

12:45-77.
73. Zakeri, Z., Curto, M., Hoover, D., Wightman, K., Englehardt, J., Smith, F.F., Kierszenbaum, A.L., Gleeson, T. and Tenniswood, M. Developmental expression of the S35-S45/SGP-2/TRPM-2 gene in rat testis and epididymis. Molec. Reprod. Devel. 1992; 33:373-384.
74. Smith, F.F., Mertz, J.R., Krebs, I., Tres, L.L., Chae, C.-B., Zakeri, Z., Englehardt, J., Hoover, D., Tenniswood, M. and Kierzenbaum, A.L. Rat Sertoli and spermatogenic cells express a similar gene and its product is antigenically related to an outer dense fiber-associated protein. Mol. Reprod. Devel. 1992; 33:363-372.
75. O'Brien, D.A., Gabel, C.A., Welch, J.E. and Eddy, E.M. Mannose 6-phosphate receptors: Potential mediators of germ cell-Sertoli cell interactions. Ann. N.Y. Acad. Sci. 1991; 637:327-339.
76. Sensibar, J.A., Qian, Y., Griswold, M.D., Sylvester, S.R., Bardin, C.W., Cheng, C.Y. and Lee, C. Localization and molecular heterogeneity of sulfated glycoprotein-2 clusterin among ventral prostate, seminal vesicle, testis and epididymis of rats. Biol.Reprod. 1993; 49:233-242.
77. Fritz, I.B. and Burdzy, K. Novel action of carnitine: inhibition of aggregation of dispersed cells elicited by clusterin in vitro. J. Cell Physiol. 1989; 140:18-28.
78. Cheng, C.Y. and Bardin, C.S. Rat testicular testibumin is a protein responsive to follicle stimulating hormone and testosterone that shares immunodeterminants with albumin. Biochemistry 1986; 25:5276-5288.
79. Cheng, C.Y., Mather, J.P., Byer, A.L. and Bardin, C.W. Identification of hormonally responsive proteins in primary Sertoli cell culture medium by high performance liquid chromatography. Endocrinol. 1986; 118:480-488.
80. Eisel, U., Reynolds, K., Riddick, M., Zimmer, A., Niemann, H. and Zimmer, A. Tetanus toxin light chain expression in Sertoli cells of transgenic mice causes alterations of the actin cytoskeleton and disrupts spermatogenesis. EMBO J. 1993; 12:3365-3372.
81. Klys, H.S., Whillis, D., Howard, G. and Harrison, D.J. Glutathione S-transferase expression in the human testis and testicular germ cell neoplasia. Br. J. Cancer 1992; 66:589-593.
82. Lamsan-Casalotti, S., Onada, M., Papadopoulos, V. and Dym, M. Developmental expression of GTP binding proteins in rat testis. J. Reprod. Fertil. 1993; 99:487-495.
83. Madden, S.L., Cook, D.M., Morris, J.F., Gashler, A., Sukhatme, V.P. and Rauscher, F.J. Transcriptional repression mediated by the WT1 Wilms' tumor gene product. Science 1991; 253:1550-1553.
84. Pritchard-Jones, K., Fleming, S., Davidson, D., Bickmore, W., Porteous, D., Gosden, C., Bard, J., Buckler, A., Pelletier, J., Houseman, D., van Heyningen, V. and Hastie, N. The candidate Wilms' tumor gene is involved in genitourinary development. Nature 1990; 346:194-197.
85. Pelletier, J., Schalling, M., Buckler, A.J., Roger, A., Haber, D.A. and Housman, D. Expression of the Wilms' tumor gene WT1 in the murine urogenital system. Genes Dev. 1991; 5:1345-1346.
86. Pelliniemi, L.J., Frojdman, K. and Paranko, J. Cell biology of testicular development. In: deKretser, D., ed. Molecular Biology of the Male Reproductive System. NY: Acad.Press, 1993:21-66.
87. Hardy, M.P., Gelber, S.J., Zhou, Z., Penning, T.M., Ricigliano, J.W.,

Ganjam, V.K., Nonneman, D. and Ewing, L.L. Hormonal control of Leydig cell differentiation. Ann. N.Y. Acad. Sci. 1991; 637:152-163.

88. Ying, S.-Y. and Murata, T. Molecular biology of the gonadal proteins-inhibin, activin, and follistatin. In: Gwatkin, R.G.B. ed. Genes in Mammalian Reproduction. New York: Wiley-Liss, 1993:207-228.
89. Choongkittaworn, N.M., Kim, K.H. Danner, D.B. and Griswold, M.D. Expression of prohibitin in rat seminiferous epthelium. Biol. Reprod. 1993; 49:300-310.
90. Schleicher, R.L., Zheng, M. and Zhang, M. Immunocytochemical localization of endogenous synthesis of apolipoprotein E in testicular Leydig cells. Biol. Reprod. 1993; 48:313-324.
91. Moore, A., Chen, C.-L. C., Davis, J.R.E. and Morris, I.D. Insulin-like growth factor-I mRNA expression in the interstitial cells of the rat testis. J. Molec. Endocrinol. 1993; 11:319-324.
92. Li, H., Risbreiger, G.P. and Clements, J.A. Pro-opiomelanocortin POMC gene expression, as identified by in situ hybridization, in purified populations of interstitial macrophages and Leydig cells of the adult rat testis. Reprod. Fertil. Develop. 1993; 5:545-554.
93. Gnessi, L., Emidi, A., Farini, D., Scarpa, S., Modesti, A., Ciampani, T., Silvestroni, L. and Spera, G. Rat Leydig cells bind platelet-derived growth factor through specific receptors and produce platelet-derived growth factor-like molecules. Endocrinol. 1992; 130:2219-2224.
94. Adham, I.M., Burkhardt, E., Benahmed, M. and Engel, W. Cloning of a cDNA for a novel insulin-like peptide of the testicular Leydig cells. J. Biol. Chem. 1993; 268:26668-26672.
95. Skinner, M.K. and Fritz, I.B. Identification of a non-mitogenic paracrine factor involved in mesenchymal-epithelial interaction between testicular peritubular cells and Sertoli cells. Mol. Cell. Endocrinol. 1986; 44:86-97.
96. Norton, J.N. and Skinner, M.K. Regulation of Sertoli cell differentiation by the testicular paracrine factor PModS: Potential role of immediate-early genes. Molec. Endocrinol. 1992; 6:2018-2026.
97. Skinner, M.K., Fetterolf, P.M. and Anthony, C.T. Purification of the paracrine factor, P-Mod-S, produced by testicular peritubular cells that modulates Sertoli cell function. J. Biol. Chem. 1988; 263:2884-2890.
98. Verhoeven, G. and Cailleau, J.A. Testicular peritubular cells secrete a protein under androgen control that inhibits induction of aromatase activity in Sertoli cells. Endocrinol. 1988; 123:2100-2110.
99. Hutson, J.C. and Stocco, C.M. Peritubular cell influence on the efficiency of androgen-binding protein secretion by Sertoli cells in culture. Endocrinol. 1981; 108:1362-1368.
100. Fritz, I.B., Tung, P.S. and Ailenberg, M. Role of interactions between Sertoli cells and peritubular cells in the control of spermatogenesis: Controls of protease activities in the seminiferous tubule. In: Baccetti, B. ed. Comparative Spermatology 20 Years After. NY: Raven, 1991:661-663.
101. Hettle, A.J., Balekjian, E., Tung, P.S. and Fritz, I.B. Rat testicular peritubular cells in culture secrete an inhibitor of plasminogen activator activity. Biol. Reprod. 1988; 38:359-371.
102. Ailenberg, M. and Fritz, I.B. Control of levels of plasminogen activator activity by Sertoli cells maintained in a two-chambered assembly. Endocrinol. 1988; 122:2613-2618.

103. Skinner, M.K., Takacs, K. and Coffey, R.J. Cellular localization of transforming growth factor-α gene expression and action in the seminiferous tubule: Peritubular cell-Sertoli cell interactions. Endocrinol. 1989; 124:845-854.
104. Skinner, M.K. and Moses, H.L. Transforming growth factor-β gene expression and action in the seminiferous tubule: peritubular cell-Sertoli cell interactions. Mol. Endocrinol. 1989; 3:625-634.
105. Tres, L.L., Smith, E.P. and Kierszenbaum, A.L. Immunoreactive sites and accumulation of somatomedin-C in rat Sertoli-spermatogenic cell co-cultures. Exp. Cell Res. 1986; 162:33-50.

CHAPTER 5

Intercellular Signaling Systems

INTRODUCTION

Coordination of spermatogenesis involves a large number of signaling pathways. These include long-range signals transmitted through the blood stream (endocrine hormonal mechanisms) and short-range signals transmitted between different cells types (paracrine mechanisms) or between identical cell types (autocrine mechanisms).

The major hormonal control mechanisms have been known for some time. These endocrine signals originate in the brain. Gonadotropin releasing hormone (GnRH), a decapeptide from the hypothalamus (Fig. 5.1),[1] stimulates the release of follicle-stimulating hormone (FSH) and luteinizing hormone (LH), complex related glycoproteins (Fig. 5.2)[2,3] from the anterior pituitary.[4] The effects of GnRH agonists and antagonists are discussed in chapter 6. FSH and LH in turn act primarily upon testicular Sertoli and Leydig cells, respectively. The most important Leydig cell product is testosterone (Fig 5.3) for which receptors exist on Sertoli, Leydig and peritubular cells. Recent work has focused on isolating and characterizing receptor genes and proteins.

Recently much attention has also been paid to the role of various growth factors in the testis. Growth factors are usually proteins present in low concentrations that promote and sometimes inhibit cell proliferation. They have broad or narrow cell-type range and act through specific receptors. In addition to influencing cell division, they may also affect differentiation and survival of cells. Spermatogenesis, which exhibits carefully controlled cell proliferation and extreme differentiation might therefore be expected to be regulated by growth factors.

Many possibilities exist for interactions among cells in the testis mediated by growth factors, some ubiquitous, some unique to the testis. Many broad specificity growth factors are made by the three major somatic cell types of the gonad: Sertoli, Leydig and peritubular cells. Since these cells can also respond to a range of factors, it not usually clear whether paracrine or autocrine mechanisms are operative. A full account pyro-Glu-His-Trp-Ser-Tyr-Gly-Leu-Arg-Pro-Gly-NH2

Fig. 5.1. Sequence of the decapeptide gonadotropin releasing hormone from porcine hypothalamus. Reproduced with permission from Russell et al. Histological and histopathological evaluation of the testis. 1990[1] ©Cache River Press

α-Subunit of Bovine Follicle Stimulating Hormone (FSH) or Luteinizing Hormone (LH)

Phe-Pro-Asp-Gly-Glu-Phe-Thr-Met-Gln-Gly-Cys-Pro-Gln-
Cys-Lys-Leu-Lys-Glu-Asn-Lys-Tyr-Phe-Ser-Lys-Pro-Asx-
Ala-Pro-Ile-Tyr-Gln-Cys-Met-Gly-Cys-Cys-Phe-Ser-Arg-
Ala-Tyr-Pro-Thr-Pro-Ala-Arg-Ser-Lys-Lys-Thr-Met-Leu-
Val-Pro-Lys-Asn-Ile-Thr-Ser-Glx-Ala-Thr-Cys-Cys-Val-
Ala-Lys-Ala-Phe-Thr-Lys-Ala-Thr-Val-Met-Gly-Asn-Val-
Arg-Val-Glx-Asn-His-Thr-Glu-Cys-His-Cys-Ser-Thr-Cys-
Tyr-Tyr-His-Lys-Ser-COOH

β-Subunit of Bovine LH

Ser-Arg-Gly-Pro-Leu-Arg-Pro-Leu-Cys-Gln-Pro-Ile-Asn-
Ala-Thr-Leu-Ala-Ala-Glu-Lys-Glu-Ala-Cys-Pro-Val-Cys-
Ile-Thr-Phe-Thr-Thr-Ser-Ile-Cys-Ala-Gly-Tyr-Cys-Pro-
Ser-Met-Lys-Arg-Val-Leu-Pro-Val-Ile-Leu-Pro-Pro-Met-
Pro-Gln-Arg-Val-Cys-Thr-Tyr-His-Glu-Leu-Arg-Phe-Ala-
Ser-Val-Arg-Leu-Pro-Gly-Cys-Pro-Pro-Gly-Val-Asp-Pro-
Met-Val-Ser-Phe-Pro-Val-Ala-Leu-Ser-Cys-His-Cys-Gly-
Pro-Cys-Arg-Leu-Ser-Ser-Thr-Asp-Cys-Gly-Pro-Gly-Arg-
Thr-Glx-Pro-Leu-Ala-Cys-Asx-His-Pro-Pro-Leu-Pro-Asp-
Ile-Leu-COOH

β-Subunit of Human FSH

Asn-Ser-Cys-Glu-Leu-Thr-Asn-Ile-Thr-Ile-Ala-Ile-Glu-
Lys-Glu-Glu-Cys-Arg-Phe-Cys-Leu-Thr-Ile-Asn-Thr-Thr-
Trp-Cys-Ala-Gly-Tyr-Cys-Tyr-Thr-Arg-Asp-Leu-Val-Tyr-
Lys-Asn-Pro-Ala-Arg-Pro-Lys-Ile-Gln-Lys-Thr-Cys-Thr-
Phe-Lys-Glu-Leu-Val-Tyr-Glu-Thr-Val-Arg-Val-Pro-Gly-
Cys-Ala-His-His-Ala-Asp-Ser-Leu-Tyr-Thr-Tyr-Pro-Val-
Ala-Thr-Gln-Cys-His-Cys-Gly-Lys-Cys-Asp-Ser-Asp-Ser-
Thr-Asp-Cys-Thr-Val-Arg-Gly-Leu-Gly-Pro-Ser-Tyr-Cys-
Ser-Phe-Gly-Glu-Met-Lys-Gln-Tyr-Pro-Thr-Ala-Leu-Ser-
Tyr-COOH

Fig. 5.2. Sequences of subunits of the bovine and human pituitary gonadotropins FSH and LH.[2,3]

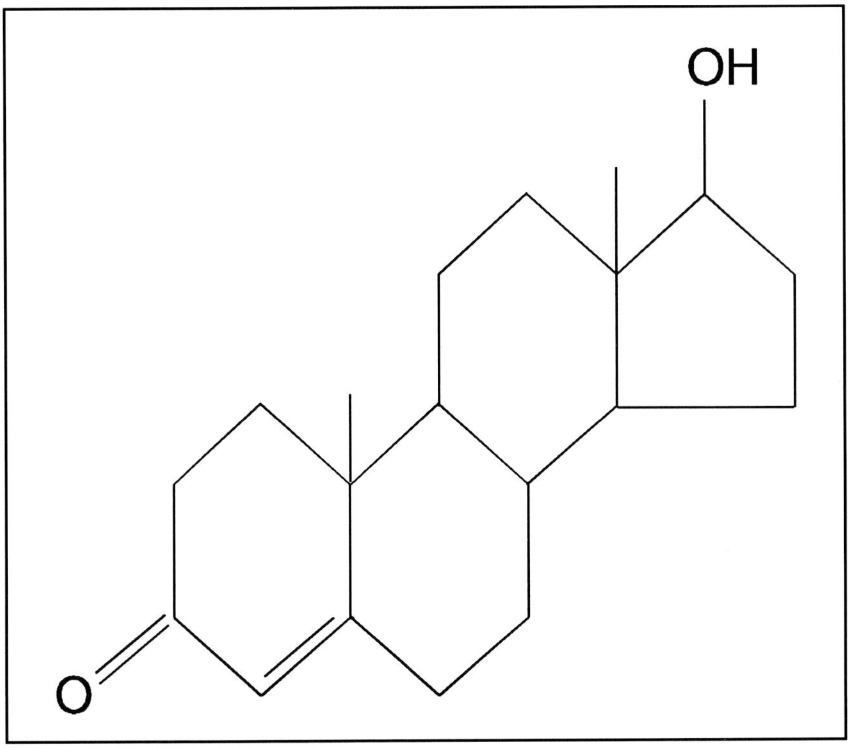

Fig. 5.3. Structure of testosterone.

of the orchestration of these components during spermatogenesis remains to be elaborated. Below is a description of the important hormones and growth factors which have received attention as actual or potential intercellular signaling molecules operating during spermatogenesis. Discussion of characterized interactions are deferred to chapter 6.

HORMONES AND HORMONE RECEPTORS

In hypophysectomized rats, only about half of the Type A spermatogonia complete mitotic divisions to form Type B spermatogonia. All Type B spermatogonia continue to proliferate, but primary spermatocytes in meiotic prophase I are lost and spermatids degenerate at Steps VIII-IX.[5] These and many other observations suggest several distinct blocks to spermatogenesis occur in the absence of pituitary hormones. Replacement therapy of hypophysectomized animals with the gonadotropins FSH and LH have helped to elucidate the roles played by these hormones in spermatogenesis. For general reviews, see refs. 4, 6, 7, 8 and 9.

Follicle stimulating hormone (FSH), also known as *follitropin,* is part of a family of glycoproteins which is produced in the pituitary or placenta including LH, thyroid stimulating hormone or thyrotropin (TSH), and human or equine chorionic gonadotropins (hCG, eCG). They are all heterodimers with common α-subunits and similarity in their hormone-specific β-subunits (Fig. 5.2).[4,10] FSH acts through a signal transduction pathway in which the FSH receptor is coupled via a G protein to adenylate cyclase resulting in increased intracellular cyclic AMP (cAMP) in Sertoli cells.[11] Increased cAMP levels activate genes for androgen-binding protein[12] and transferrin[13] and for the transcription factors c-*fos* and *jun*-B.[14]

High affinity, low capacity *FSH receptors* are found on Sertoli cell membranes,[15] but not on any other testicular cells. Rat, human, monkey, and sheep receptor genes have been cloned and predict very similar proteins.[12,16-18] Each has a large extracellular domain, seven transmembrane spanning hydrophobic regions, and a short cytoplasmic tail shared in common with other members of the receptor family including LH/CG and TSH receptors. All three FSH receptors have potential N-linked glycosylation sites and a motif characteristic of a G-protein coupled receptor family. The sheep FSH-R is a protein of apparent 75 kDa molecular weight containing 678 amino acids.[17] Transfection of fetal kidney cells with sheep receptor cDNA confers receptor expression on the plasma membrane and high affinity and high specificity FSH binding.[17]

The sheep protein is encoded by two major transcripts of 2.6 and 6.7 kb.[17] Multiple transcripts of FSH receptor from rat and mouse have also been demonstrated.[19] A major rat FSH receptor transcript of 2.6 kb and a minor 4.5 kb transcript were reported in cultured Sertoli cells and in whole testis. The mRNA shows a 5-fold variation during the seminiferous cycle, reaching a maximum at Stages XII to II.[20] This is the same time as the maximum for binding of FSH[21] and corresponds to the time of spermatogonial mitoses and spermatocyte meiosis. However, other reports put maximal FSH receptor expression in rats at Stages IX-X.[22] FSH receptor mRNAs of 2.4 and 5.5 kb in rat were reported to decrease between days 7 and 14, then remained constant.[23] In addition to the full length sheep FSH receptor,[17] two alternatively spliced forms of testicular FSH receptor mRNA have been cloned.[24] One of 727 bp codes for a putative soluble receptor of 134 amino acids, the second of 1.1 kb encodes a 259-amino acid form with a predicted single transmembrane domain. Both lack the G protein coupling domains. Incomplete or alternatively spliced transcripts have also been reported for the human FSH receptor.[25-27] Alternate forms of FSH-receptor may modulate responses of different target cells, but this remains to be demonstrated. In addition, regulation of FSH receptor number during spermatogenesis may vary with the strain of mouse.[28]

Luteinizing hormone (LH) triggers Leydig cells to produce testosterone (Fig. 5.3).[29] Long-term suppression of LH leads to loss of sperm which can be reversed completely by testosterone therapy.[30] Human chorionic gonadotropin from the placenta, hCG, stimulates testosterone production and increases cAMP levels.[31] Evidence for and against cAMP as a sole or obligatory second messenger in this system have been recently reviewed.[8] Other candidates suggested to be involved in signal transduction are Ca^{++}, protein kinase C, Cl^- channels, cGMP and tyrosine kinases.

The *LH/CG (chorionic gonadotropin) receptor* is also a G-protein coupled receptor.[32] It has been cloned from rat[33] and pig.[43] Levels of receptor remain constant after puberty in the mouse.[28] In rat, the receptor is a 93 kDa protein with seven transmembrane spanning regions and a very long extracellular domain with many leucine-rich repeats.[35] The gene is 70 kb long with 10 exons.[36,37] The promoter structure yields few clues about specific regulatory factors. Multiple transcripts in Leydig cells suggest alternative or incomplete splicing.[38] One encodes a truncated form corresponding to the extracellular domain suggesting a possible secreted form.[39-41] LH receptors lacking the cytoplasmic domain transfected into kidney cells bind hCG similarly to the complete receptor.[41] Ligand binding occurs in the extracellular domain but a binding site in the transmembrane domain may also be involved.[42] Some responses such as the LH stimulation of cAMP and the internalization of the ligand are increased after removal of the cytoplasmic portion of the receptor.[43,44]

Just as gonadotropins from the pituitary regulate the testis, inhibins, activin and follistatin from the testis feed back to regulate the pituitary (Fig. 5.4). Inhibins and activins are related molecules and can bind to follistatin.

Inhibin is a glycoprotein that in males is secreted by Sertoli cells. Inhibin binds in the pituitary and suppresses FSH but not LH synthesis

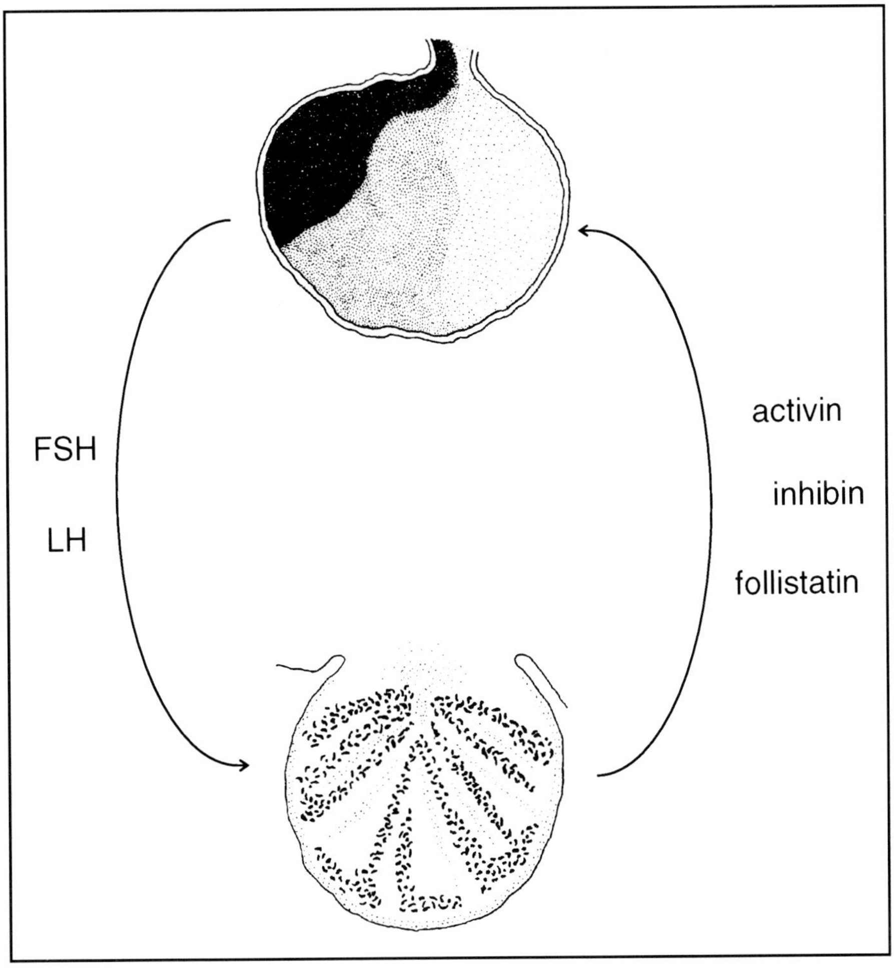

Fig. 5.4. Some interactions of the pituitary and the testis. (Drawing by J. Poccia.)

and release.[45,46] It has also been reported to suppress spermatogonial proliferation.[46a] Multiple inhibin activities have been found: one of 5-20 kDa in seminal plasm and one of 40-70 kDa in gonadal extracts.[47] As isolated from cow follicular fluid, inhibin is a 32 kDa protein consisting of two polypeptide chains linked by disulfide bonds.[46] A common α-subunit is joined to either β_A or β_B.[48] It is similar to TGF-β in amino acid sequence[49] and its superfamily includes anti-Mullerian factor.[50] For extensive reviews, see refs. 15 and 50-53.

In the pig, two forms of inhibin, A and B, each 32 kDa, are found. The 32 kDa dimer is composed of an 18 kDa α- and 14 kDa β-subunit. The common α-subunit is 134 amino acids, the β-subunit A is 116 and β-subunit B is 115 amino acids. The unprocessed β-subunit joins to various α-subunits by disulfide bonds. The mature α- and β- subunits are made from precursors, encoded by a single gene. Larger forms of inhibin probably result from unprocessed molecules. The human α precursor is 364 amino acids long. The human β precursors are 424 amino acids (βA) and 407 amino acids (βB).[49]

A single mRNA of 1.6 kb encoding the α-subunit is found in rat Sertoli cells. Inhibin β_A and β_B are encoded by two and three mRNAs respectively. Inhibin β_A mRNA in Sertoli cells accumulates at Stage VIII and is maximally expressed at Stages IX-XI.[54] Inhibin β_B mRNA is maximally expressed at Stages XIII-III,[54] similar to inhibin α mRNA. Inhibin β_A and β_B mRNAs increase around germ cells in which receptor is accumulating.[54] Mice with null mutations which abolish α-subunit activity develop gonadal tumors suggesting a potential tumor suppressor role.[55]

Curiously, a 19 kDa nuclear protein previously reported to be a sperm-specific histone is related to α-inhibins and appears to be a proteolytic product of a 52 kDa protein called semenogelin excreted from cells of the seminal vesicle.[56]

Activins are 28 kDa dimers of β-subunits of inhibin which stimulate the synthesis and release of FSH.[50,57] They are secreted by Sertoli and many other cell types[58] and exist as either homodimers of β_A (activin A) or β_B (activin B) or as heterodimers of β_A and β_B (activin AB). Inhibin and activin are part of the protein family which includes TGF-βs, AMH and other gene products with roles in differentiation.

Activin receptors are present on spermatogonia and some primary spermatocytes.[59] Activin receptors have been cloned from mouse and rat.[57,60] A 513-amino acid precursor molecule matures to a 494-amino acid protein.[60] The 56 kDa protein has a 116-amino acid N-terminal extracellular ligand binding domain which is cysteine-rich and a 352-amino acid intracellular serine-threonine kinase signaling domain. Two potential N-glycosylation sites are found in the extracellular domain and several phosphorylation sites are present in the intracellular domain. Two activin receptor mRNAs, 6 and 4 kb, are present in testes of immature and mature rats.[61] An increase in the 4 kb mRNA at 21-28 days postnatally suggests germ cell expression. The 4 kb mRNA is detected in isolated pachytene spermatocytes and round but not elongating spermatids.[57,62] Sertoli cells express both messages; levels are low in Leydig cells. Activin A receptor mRNA is maximal in spermatocytes at Stages XIII-XIV and in early round spermatids at Stages I-IV.[54] Activin receptor mRNA is found in Type A spermatogonia at Stages IX-XI.[63] Two activin receptor mRNAs are expressed in rat testis, a 6 kb and a 3 kb form that increase from day 7 to 35.[23]

In addition to regulating pituitary activities, inhibins and activins may have short range effects. Both affect steroidogenesis. Germ line cells can also respond to activin. *Inhibin A receptors* are found on spermatogonia, primary spermatocytes and early spermatids.[59] Activin stimulates proliferation of spermatogonia in coculture with Sertoli cells.[59] Activins may also regulate Leydig steroidogenesis and spermatogenesis. Inhibin βA has been localized to the plasma membrane of spermatids.[64]

Follistatins are glycosylated monomeric proteins of at least three forms: 32, 35 and 37 kDa. They inhibit the release of FSH by cultured pituitary cells and modulate activin activity.[54] Follistatin binds to activin A or the β subunit of inhibin.[65] Follistatins contain four domains: a signal peptide and three cysteine-rich domains similar to one another. Two precursor proteins of 344 and 317 amino acids have been identified. Two cDNAs produced by alternative splicing encode the two precursors. Follistatin mRNA accumulates at Stage VIII and is maximally expressed at Stages IX-XI.[54]

The testis secretes androgens such as *testosterone*, *androsterone* and *dehydroepiandrosterone* (predominantly from the Leydig cells) and estrogens (from Leydig and Sertoli cells).[9] Androstenedione is synthesized from progesterone which is made from cholesterol. Androstenedione is reversibly converted to testosterone by 17β-hydroxysteroid dehydrogenase (OHSD).[9,66] Both testosterone and androstenedione are converted to the estrogens *estradiol* and *estrone* respectively by an aromatase. Two genes for OHSD have been cloned.[67] Type II is expressed exclusively in testis and adrenals and type I predominantly in the placenta. Regulation of steroid pathways in Leydig cells has been recently reviewed.[9]

Testosterone is the most important steroid produced by the Leydig cells in response to LH through a cAMP signaling pathway (Fig. 5.3). It is needed to initiate spermatogenesis at puberty and maintain it in the adult. Testosterone is required for completion of meiosis and for differentiation of spermatids. The concentration of testosterone in the testis is much higher than needed to saturate its receptors in adult rat testis.[68] Dihydrotestosterone can substitute for testosterone but its lower concentration in the testis suggests testosterone is the active agent in vivo.[30] Testosterone may be required for maintenance of Sertoli-Sertoli cell junctions and thus for the integrity of the blood-testis barrier.[69] For reviews see refs. 7 and 9.

Chimeric mice exhibit normal spermatogenesis if germ cells lack functional *androgen receptors*[70] indicating that somatic cells are the targets of androgens. Receptors have been localized to Sertoli, Leydig and peritubular cells but not germ cells.[71-75] Sertoli cells have high affinity, low capacity androgen receptors in both cytoplasm and nucleus.[71,72] Cytoplasmic androgen receptor binds with high affinity to testosterone ($K_d = 7.6 \times 10^{-10}$), and is clearly different from androgen binding protein. The nuclear receptor has a K_d of 2.5×10^{-9} M. Androgen receptor genes have been cloned from human and rat.[76,77] The human and rat receptors show hormone and DNA binding domains typical of steroid hormone receptors: a variable N-terminal region functioning in gene expression, a highly conserved DNA binding region with two zinc finger motifs, and a less well-conserved C-terminal region which binds hormone and is implicated in dimerization.[78] Co-transfection of the receptor and a reporter gene shows that the receptor acts as a ligand-activated transcription factor.[79] Androgen receptor mRNA shows a 2-fold variation in the seminiferous cycle with a maximum at Stages XIII to I, the same as does FSH receptor mRNA.[20]

Several human mutations have been associated with androgen insensitivity syndrome. Complete lack of androgen receptor function results in testicular feminization including external female genitalia and intra-abdominal testes in XY individuals. Over 10 mutations leading to complete androgen insensitivity have been described. In one, a single methionine 786 is changed to valine in the steroid binding domain.[80] In another, a single arginine is changed to histidine at amino acid 614 in the second zinc finger region of a DNA binding domain.[81] Partial androgen responsiveness results in incomplete testicular feminization in males. One of these results from a change of glycine to valine at position 743 in the steroid binding domain.[82] Several other mutations have been recently reported affecting the receptor.[80,83-95]

Retinoic acid is produced by oxidation of retinol. Retinoids are steroid hormones made from vitamin A that can diffuse across the plasma membrane like testosterone. Retinoic acid restores spermatogenesis in vitamin A deficient rats.[96] *Retinoic acid receptors* (*RARs*) are DNA binding proteins similar to steroid hormone receptors. The mRNAs for RARα and RARβ receptors are maximal in Stages VII-XI which correlates with the stages blocked by vitamin A deficiency.[20,97] A 2.9 kb mRNA for RARβ and a 2.7 kb mRNA for RARα are expressed only in Sertoli cells. A 3.4 kb mRNA is expressed in spermatocytes as well as Sertoli cells. Retinoids induce RARα mRNA.[97,98] The gene for RARα was disrupted in transgenic mice giving rise to degeneration of testis and was lethal at an early postnatal stage.[99] Degeneration of the germinal epithelium in RARα-null homozygotes was similar to the effects of vitamin A deficiency in mice suggesting retinoic acid not retinol accounts for the dietary effects. If RARα1, the ubiquitously expressed isoform, is specifically disrupted, mice do not show abnormalities, implying that the α1 and α2 forms are redundant. Cellular *retinol binding protein* is found in large amounts in late spermatocytes and spermatids[100] and is expressed cyclically by rat Sertoli cells before late germ cells are present.[101] It may serve to transport retinol to germ cells.[102]

Insulin, a polypeptide hormone, stimulates the formation of lactate from glucose in Sertoli cells.[103] Lactate is the major energy source for spermatogenesis.[103] Insulin is a member of a related family comprised of insulin, relaxin and insulin-like growth factors. A novel member of the insulin-like peptide family, *Ley I-L,* is expressed exclusively in Leydig cells.[104] It is made as a preproprotein of 131 amino acids with a 24-amino acid signal peptide. Predicted proteolytic processing produces a 32-amino acid B-chain and a 26-amino acid long A chain.

GROWTH FACTORS AND GROWTH FACTOR RECEPTORS

Because gonadotropins are of relatively constant concentrations in the circulation, growth factors have been postulated to control and regulate Sertoli and Leydig cell development. These may exert effects on the germ cells as well. Some are only partially characterized. Others have been cloned or have been well-investigated in other tissues. For further reviews, see refs. 105, 106.

Broad specificity growth factors found in many tissues as well as in the testis include transforming growth factors TGF-α and TGF-β, epidermal growth factor EGF, basic fibroblast growth factor bFGF, insulin-like growth factor IGF, and platelet-derived growth factor PDGF.

Transforming growth factor-α (TGF-α) is a mitogen, a member of a family of epidermal-like growth factors which mimic EGF by binding to the EGF receptor.[107] TGF-α is expressed by Sertoli and peritubular cells in culture[108] and has been immunolocalized to Leydig cells and to peritubular cells.[109] Its mRNA is expressed maximally in pubertal peritubular cells, but is constant during development in Sertoli cells.[107] TGF-α is made as a transmembrane precursor molecule which can be cleaved to its mature form or can act on adjacent cells.[110]

Transforming growth factor-β (TGF-β)[111] is a potent growth inhibitor and inducer of differentiation. In the rat, peritubular cells, Leydig cells, Sertoli cells and germ cells express two different gene products: TGF-β1 and TGF-β2.[112,113] TGF-β1 and TGF-β2 mRNAs are highest in immature testis, decreasing at puberty. A TGF-β3 mRNA is expressed by germ cells.[114] TGF-β3 mRNA peaks coincidentally with the initiation of spermatogenesis.[111] Germ cells make a unique TGF-β1 transcript.[114] Immunochemical localization reveals TGF-β1 predominantly in spermatocytes and early round spermatids which then declines whereas TGF-β2 appears in later spermatids.[112] TGF-β1 is found in Sertoli cells throughout development, but TGF-β2 peaks in fetal Sertoli cells, then declines. In fetal Leydig cells, TGF-β1 and TGF-β2 are present, but TGF-β2 declines after birth. Little TGF-β was found in adult Leydig cells.[112]

An *epidermal growth factor (EGF)*-like molecule of 2-8 kDa has been isolated from Sertoli cell conditioned medium.[115] EGF is usually a 53-amino acid protein processed proteolytically from a precursor.[116] EGF alters androgen binding protein, lactate, ornithine decarboxylase, inhibin production and transferrin secretion of Sertoli cells and influences the production of androgens by Leydig cells.[116] Lowering EGF levels in mice interferes with spermatogenesis while having little effect on FSH or testosterone levels, and the effect can be reversed upon addition of EGF.[117] *EGF receptors,* 170 kDa transmembrane proteins, have been reported to be immunochemically localized variously in Leydig, Sertoli and peritubular cells,[118] in Sertoli and Leydig but not peritubular cells,[119,120] or in peritubular cells only.[108] Gene expression is high in peritubular cells, low in Sertoli cells. High affinity receptors have been found in peritubular cells, none on Sertoli cells. Low levels of EGF receptors have been reported on pachytene spermatocytes and round spermatids, but the protein has not yet been identified. EGF-R is also a receptor for

TGF-α.[121] Functional EGF receptors are also found on the post-acrosomal region of human sperm where they have been suggested to have a role in fertilization.[122]

Basic fibroblast growth factor (*bFGF*) is a 30 kDa protein[123,124] whose mRNA is expressed in pachytene spermatocytes.[123] It has been isolated as three peptides of 24, 27 and 30 kDa from germ cell conditioned medium.[124] Germ cells seem to be the main source of bFGF in the testis, although it has been immunolocalized to Leydig cells as well.[124] Two forms of *bFGF receptors* of 120 kDa and 143 kDa are found in round and elongated spermatids.[124] Only the 120 kDa form is present in Sertoli cells.[124] Receptors are maximally expressed at Stages I-VIII.[124]

Platelet derived growth factor (PDGF)-like molecules of about 30 kDa have been reported to be produced by Leydig cells.[125] They consist of all three combinations of disulfide-linked dimers of closely related A and B subunits. The B chain is encoded by the *c-sis* proto-oncogene. *PDGF-receptors* have been demonstrated on adult rat Leydig cells[125] and on peritubular cells.[126]

Somatomedin C or *insulin-like growth factor (IGF-I)* is expressed by Sertoli and peritubular cells in the testis, Leydig and other interstitial cells and non-testicular sources as well.[115,127,128] Somatomedin-like growth factors act as mitogens on spermatogonia.[129] IGF-I is an 8 kDa protein.[129] Most testicular IGF-I is found in Sertoli cells and primary spermatocytes.[130] IGF-I mRNA is detected in Leydig cells and in populations of interstitial cells depleted of Leydig cells.[128] In humans, IGF-I mRNA was not detected, but IGF-II mRNA was abundant in blood vessels and peritubular cells.[131] *IGF-I receptors* are found on secondary spermatocytes and early spermatids. Sertoli cells have a high affinity receptor.[129,132-134] Leydig cells are also reported to have IGF-I receptors.[135] IGF-I and IGF-II receptor mRNAs were abundant in germinal epithelium. *IGF binding proteins* are produced by Sertoli cells (IGFBP-3) and peritubular cells (IGFBP-2).[136,137] In humans, IGFBP-1 mRNA was not detected, but IGFBP-2 mRNA was expressed in Sertoli and Leydig cells, IGFBP-4 in Leydig cells and interstitial connective tissue, IGFBP-5 at low levels in Leydig cells, and IGFBP-6 in some interstitial cells and peritubular cells.[131]

Prohibitin is an intracellular negative growth regulator of 30 kDa expressed in a wide variety of tissues. Two transcripts of 1.9 and 1.2 kb are present in Sertoli cells.[138] The protein is expressed constitutively in Sertoli and Leydig cells of the adult rat.

Other growth factors believed to be of narrower specificity found in the testis include interleukins and nerve growth factors. *Interleukin-1* is produced stage-specifically in testis, correlating with spermatogonial proliferation.[139-141] *Interleukin-6* increases following Stages VII- VIII.[142] *Interleukin receptors* have been reported in germ cells.[142a]

Nerve growth factor (*β-NGF*) is a 118-amino acid long protein which belongs to a family of related proteins expressed in brain. It is also expressed in spermatogenic cells and stimulates DNA synthesis in meiotic cells.[143] Low affinity *NGF receptors* are found on Sertoli cells with a K_d of 10^{-9}-10^{-10}.[144] They have a cytoplasmic domain, a membrane-spanning region and a cysteine-rich extracellular domain. The low affinity *NGF receptor* from liver is a 75 kDa glycoprotein which complexes with a 140 kDa peptide encoded by the *c-trk* proto-oncogene, a tyrosine kinase receptor. Complex formation converts the 75 kDa protein to the high affinity NGF-R when co-expressed.[145] Only high affinity receptors mediate activity. They have a K_d of 10^{-10}-10^{-11}. *Trk* is expressed at all stages of the spermatogenic cycle in rat Sertoli cells.[143,146] Low affinity receptor mRNA is maximal in Stages VI-VIII during the beginning of meiosis and the protein is found on the plasma membrane of Sertoli cells.[143]

Some testicular factors appear to be unique to the testis or specifically involved in its regulation. For example, Sertoli cells make several factors that regulate Leydig cells.

Leydig cell stimulatory factor is a 10 kDa protein which stimulates steroidogenesis in Leydig cells and has been purified from cultured Sertoli cell medium.[147] Sertoli cells also secrete *Leydig cell stimulator*,[31] a 17 kDa molecule immunologically different from LH. In contrast to LH, it stimulates testosterone production without changing cAMP levels.[31] *Leydig cell inhibitor* inhibits cAMP levels, hCG-stimulated androgen production by Leydig cells and conversions of various precursors to testosterone.[31]

Seminiferous growth factor, SGF, enhances epithelial cell proliferation and is present in prepubertal Sertoli cells.[148,149] It is a heparin-binding growth factor presumably of the FGF family.

SUMMARY

Intercellular signaling molecules are necessary to integrate testicular function and promote spermatogenesis. Candidates for intercellular signaling molecules include hormones and growth factors and their receptors. Important hormones are gonadotropin-releasing hormone from the hypothalamus, follicle stimulating hormone and luteinizing hormone from the pituitary and testosterone from the Leydig cells. Hormones from the testis affecting pituitary function are the inhibins, activins and follistatin. Retinoic acid and insulin also control testicular function. Broad specificity growth factors of the testis include TGF-α, TGF-β, EGF, bFGF, PDGF, IGF-I and prohibitin. Narrower specificity factors include IL-1, IL-6, and β-NGF. Factors thought to be unique to the testis include Leydig cell stimulatory and inhibitory factors and seminiferous growth factor.

REFERENCES

1. Matsuo, H., Baba, Y., Nair, R.M., Ariuma, A. and Schally, A.V. Structure of the porcine LH- and FSH-releasing hormone. I. The proposed amino acid sequence. Biochem. Biophys. Res. Comm. 1971; 43:1334-1339.
2. Liu, W.-K., Nahm, H.S., Sweeney, C., Holcomb, G.N. and Ward, D.N. The primary structure of bovine luteinizing hormone Part 2. The amino-acid sequence of the reduced S carboxymethylated A subunit. J. Biol. Chem. 1972; 247:4365-4381.
3. Saxena, B.B. and Rathnam, P. Amino-acid sequence of the β subunit of follicle stimulating hormone from human pituitary glands. J. Biol. Chem. 1976; 251:993-1005.
4. Pierce, J.G. and Parsons, T.F. Glycoprotein hormones: Structure and Function. Ann. Rev. Biochem. 1981; 50:465-495.

5. Clermont, Y. and Morganthaler, H. Quantitative study of spermatogenesis in the hypophysectomized rat. Endocrinol. 1955; 57:369-382.
6. Steinberger, E. Hormonal control of mammalian spermatogenesis. Physiol. Rev. 1971; 51:1-22.
7. Weinbauer, G.F. and Nieschlag, E. Hormonal control of spermiogenesis. In: deKretser, D. ed. Molecular Biology of the Male Reproductive System. NY: Acad.Press, 1993:99-142.
8. Huhtaniemi, H. Hormonal control mechanisms of Leydig cells. In: deKretser, D. ed. Molecular Biology of the Male Reproductive System. NY: Acad. Press, 1993:383-410.
9. Hall, P. Molecular biology of testicular steroid secretion. In: deKretser, D. ed. Molecular Biology of the Male Reproductive System. NY: Acad. Press, 1993:328-382.
10. Sairam, M.R. and Manjunath, P. Studies on pituitary follitropin 11. Induction of hormonal antagonistic activity by chemical deglycosylation. Mol. Cell. Endocrinol. 1982; 28:139-150.
11. Means, A.R., Dedman, J.R., Tash, D.J., Tindall, D.J., van Sickle, M. and Welsh, M.J. Regulation of the testis Sertoli cell by follicle stimulating hormone. Ann. Rev. Physiol. 1980; 42:59-70.
12. Sprengel, R., Braun, T., Nikolics, K., Segaloff, D.L. and Seeburg, P.H. The testicular receptor for FSH. Structural and functional expression of cloned complementary DNA. Mol. Encdocrinol. 1990; 4:525-530.
13. Hall, J.N. and Skinner, M.K. Regulation of Sertoli cell function and differentiation through the actions of a testicular paracrine factor P-Mod-S. Endocrinol. 1989; 124:2711-2719.
14. Hall, S.H., Joseph, D.R., French, F.S. and Conti, M. Follicle stimulating hormone induces transient expression of the proto-oncogene *c-fos* in primary Sertoli cell cultures. Mol. Endocrinol. 1988; 2:55-61.
15. Tindall, D.J., Rowley, D.R., Murthy, L., Lipshultz, L.I. and Chang, C.H. Structure and biochemistry of the Sertoli cell. Int. Rev. Cytol. 1985; 94:127-149.
16. Minegish, T., Nakamura, K., Takadura, Y., Ibuki, Y. and Igarashi, M. Cloning and sequencing of human LH/hCG receptor cDNA. Biochem. Biophys. Res. Comm. 1990; 172:1049-1054.
17. Yarney, T.A., Sairman, M.R., Khan, H., Ravindranath, N., Payne, S. and Seidah, N.G. Molecular cloning and expression of the ovine testicular follicle stimulating hormone receptor. Molec. Cell. Endocrinol. 1993; 93:219-226.
18. Grommol, J., Dankbar, B., Sharma, R.S. and Nieschlag, E. Molecular cloning of the testicular follicle stimulating hormone receptor of the non human primate *Macaca fascicularis* and identification of multiple transcripts in the testis. Biochem. Biophys. Res. Comm. 1993; 196:1066-1072.
19. Heckert, L. and Griswold, M.D. The changing functions of follicle-stimulating hormone in the testes of prenatal, newborn, immature and adult rats. In: Hunzicker-Dunn M. and Schwartz, N.B. eds. Follicle-Stimulating Hormone-Regulation of Secretion and Molecular Mechanism of Action. New York: Springer-Verlag, 1992:237-245.
20. Linder, C.C., Heckert, L.L., Roberts, K.P., Kim, K.H. and Griswold, M.D. Expression of receptors during the cycle of the seminiferous epithelium. Ann. N.Y. Acad. Sci. 1991; 637:313-321.

21. Kangasniemi, M., Kaipia, A., Toppari, J., Perheentupa, A., Huhtaniemi, I. and Parvinen, M. Cellular regulation of follicle-stimulating hormone FSH binding in rat seminiferous tubules. J. Androl. 1990; 11:336-343.
22. Kliesch, S., Penttila, S., Gromoll, J., Saunders, P.T.K., Nieschlag, E. and Parvinen, M. FSH receptor mRNA is expressed stage-dependently during spermatogenesis. Mol. Cell. Endocrinol. 1992; 84:R45-R49.
23. Ito, I., Minegishi, T., Hasegawa, Y., Shinozaki, H., Nakamura, K., Igarashi, S., Nakamura, M., Miyamoto, K. and Ibuki, Y. Developmental changes of testicular activin and FSH receptor mRNA and plasma FSH and inhibin levels in the rat. Life Sci. 1993; 53:1299-1307.
24. Khan, H., Yarney, T.A. and Sairam, M.R. Cloning of alternatively spliced mRNA transcripts coding for variants of ovine testicular follitropin receptor lacking the G protein coupling domains. Biochem. Biophys. Res. Comm. 1993; 190:888-894.
25. Kelton, C.A., Cheng, S.V.Y., Nugent, N.P., Schweikhardt, R.L., Rosenthal, R.L., Overton, J.L., Overton, S.A., Wands, G.D., Kuzeja, J.B., Luchette, C.A. and Chappel, S.C. The cloning of human follicle stimulating hormone receptor and its expression in Cos-70, CHO and Y-1 cells. Mol. Cell. Endocrinol. 1992; 89:141-151.
26. Braun, T., Schofield, P.R. and Sprengel, R. Amino-terminal leucine-rich repeats in gonadotropin receptors determine hormone selectivity. EMBO J. 1991; 10:1885-1990.
27. Gromoll, J., Guderman, T. and Nieschclag, E. Molecular cloning of a truncated isoform of the human follicle-stimulating hormone receptor. Biochem. Biophys. Res. Comm. 1992; 188:1077-1083.
28. Shiraishi, K., Singtripop, T., Park, M.K. and Kawashima, S. Developmental changes of testicular gonadotropin receptors and serum gonadotropin levels in two strains of mice. Zool. Sci. 1993; 10:497-504.
29. Ewing, L.L. and Zirkin, B.R. Leydig cell structure and steroidogenic function. Recent Prog. Horm. Res. 1983; 39:599-635.
30. Roberts, K.P. and Zirkin, B.R. Androgen regulation of spermatogenesis in the rat. Ann. N.Y. Acad. Sci. 1991; 637:90-106.
31. Cheng, C.Y., Zwain, I., Grima, J. and Bardin, C.W. Structure and functions of Sertoli cell proteins. In: Baccetti, B. ed. Comparative Spermatology 20 Years After. NY: Raven, 1991:617-624.
32. Strosberg, A.D. Structure-function relationship of proteins belonging to the family of receptors coupled to GTP-binding proteins. Eur. J. Biochem. 1991; 196:1-10.
33. McFarland, K.C., Sprengel, R., Phillips, H.S., Kohler, M., Rosemblit, N., Nikolics, K., Segaloff, D.L. and Seeburg, P.H. Lutropin-choriogonadotropin receptor: An unusual member of the G protein-coupled receptor family. Science 1989; 245:494-499.
34. Loosfelt, H., Misrahi, M., Atger, M., Salesse, R., Thi, M.T.V.H.-L., Jolivet, A., Guiochon Mantel. A.G., Sar, S., Jallal, B., Garnier, J. and Milgrom, E. Cloning and sequencing of porcine LH-hCG receptor cDNA: Variants lacking transmembrane domain. Science 1989; 245:525-528.
35. Metsikko, M.K., Petaja-Repo, U.E., Lakkakorpi, J.T. and Rajaniemi, H.J. Structural features of the LH/GC receptor. Acta Endocrinol. 1990; 122:545-562.
36. Koo, Y.B., Ji, I., Slaughter, R.G. and Ji, T.H. Structure of the luteinzing

hormone receptor and multiple exons of the coding sequence. Endocrinol. 1991; 128:2297-2308.

37. Tsai-Morris, C.H., Buczko, E., Wang, W., Xie, X.-Z. and Dufau, M.L. Structural organization of the rat luteinizing hormone LH receptor gene. J. Biol. Chem. 1991; 266:11355-11359.
38. Hsueh, A.J.W. and LaPolt, P.S. Molecular basis of gonadotropin receptor regulation. Trends Endocrinol. Metab. 1992; 3:164-170.
39. Xie, Y.-B., Wang, H. and Segaloff, D.L. Extracellular domains of lutropin/ choriogonadotropin receptor expressed in transfected cells binds choriogonadotropin with high affinity. J. Biol. Chem. 1990; 265: 21411-21414.
40. LaPolt, P.S., Jia, X.-C., Sincich, D. and Hsueh, A.J.W. Ligand-induced down-regulation of testicular and ovarian luteinizing hormone LH receptors is preceded by tissue-specific inhibition of alternatively processed LH receptor transcripts. Mol. Endocrinol. 1991; 5:397-403.
41. Tsai-Morris, C.H., Buczko, E., Wang, W. and Dufau, M.L. Intronic nature of the rat luteinizing hormone receptor gene defines a soluble receptor subspecies with hormone binding. J. Biol. Chem. 1990; 266: 19385-19388.
42. Ji, I. and Ji, T.H. Exons 1-10 of the rat LH receptor encode a high affinity hormone binding site and exon 11 encodes a G-protein modulation and a potential second hormone binding site. Endocrinol. 1991; 128:2648-2650.
43. Rodriquez, M.C., Xie, Y.-B., Wang, H., Collison, K. and Segaloff, D.L. Effects of truncation of the cytoplasmic tail of the luteinizing hormone/ chorionic gonadotropin receptor on receptor-mediated hormone internalization. Mol. Endocrinol. 1992; 6:327-336.
44. Sanchez-Yague, J., Rodriquez, M.C., Segaloff, D.L. and Ascoli, M. Truncation of the cytoplasmic tail of the lutropin/choriogonadotropin receptor prevents agonist-induced uncoupling. J. Biol. Chem. 1992; 267:7217-7220.
45. Ling, N., Ying, S.-Y., Ueno, N., Esch, F., Denoroy, L. and Guillemin, R. Isolation and partial characterization of a Mr 32,000 protein with inhibin activity from porcine follicular fluid. Proc. Natl. Acad. Sci. USA 1985; 82:7217-7221.
46. Robertson, D.M., Foulds, L., Leversha, L., Morgan, F.J., Hearn, M.T.V., Burger, H.G., Wettenhall, R.E.H. and deKretser, D.M. Isolation of inhibin from bovine follicular fluid. Biochem. Biophys. Res. Comm. 1985; 126:220-226.

46a. van Dissel-Emiliani, F.M.F., Grootenhuis, A.J., de Jong, F.H. and de Rooij, D.G. Inhibin reduces spermatogonial numbers in testes of adult mice and Chinese hamsters. Endocrinol. 1989; 125: 1899-1903.

47. De Jong, F.H. and Robertson, D.M. Inhibin 1985. Update of action and purification. Mol. Cell. Endocrinol. 1985; 42:95-103.
48. Esch, F.S., Shimaski, S., Cooksey, K., Mercado, M., Mason, A.J., Ying, S.-Y., Ueno, N. and Ling, N. Complementary dexoxyribonucleic acid cDNA cloning and DNA sequence analysis of rat ovarian inhibins. Mol. Endocrinol. 1987; 1:388-396.
49. Mason, A.J., Hayflick, J.S., Ling, N., Esch, F., Ueno, N., Ying, S.-Y., Guillermin, R., Niall, H. and Seeburg, P.H. Complementary DNA sequences of ovarian follicular fluid inhibin show precursor structure and

homology with transforming growth factor-β. Nature 1985; 318:659-663.
50. Ying, S.-Y. and Murata, T. Molecular biology of the gonadal proteins-inhibin, activin, and follistatin. In: Gwatkin, R.G.B. ed. Genes in Mammalian Reproduction. New York: Wiley-Liss, 1993:207-228.
51. De Jong, F.H. Inhibin. Physiol. Rev. 1988; 68:555-607.
52. Vaughn, J.M., Rivier, J., Corrigan, A.Z., McClintock, R., Campen,, C.A., Jolley, D., Voglmayr, J.K., Bardin, C.W., Rivier, C. and Vale, W. Inhibin. Meth. Enzymol. 1989; 168:588-617.
53. Vale, W., Rivier, C., Hseuh, A., Campen, C., Meurnier, H., Bicsak, T., Vaughn, J., Corrigan, A., Bardin, W., Sawchecko, P., Petraglia, F., Plotsky, P., Spiess, J. and Rivier, C. Chemical and biological characterization of the inhibin family of protein hormones. Recent Prog. Horm. Res. 1988; 44:1-34.
54. Kaipia, A., Penttila, T.-L., Shimasaki, S., Ling, N. Parvinen, M. and Toppari, J. Expression of inhibin β-A and β-B, follistatin and activin-A receptor messenger ribonucleic acids in the rat seminiferous epithelium. Endocrinol. 1992; 131:2703-2710.
55. Matzuk, M.M., Finegold, M.J., Su, S.-J., Hsueh, A.J.W. and Bradley, A. α-Inhibin is a tumor suppressor gene with gonadal specificity in mice. Nature 1992; 360:313-319.
56. Zalensky, A.O., Yau, P., Breneman, J.W. and Bradbury, E.M. The abundant 19-kilodalton protein associated with human sperm nuclei that is related to seminal plasma α-inhibins. Molec. Reprod. Devel. 1993; 36:164-173.
57. deWinter, J.P., Themmen, A.P.N., Hoogerbrugge, J.W., Klaij, L.A., Grootegoed, J.A. and de Jong, F.H. Activin receptor mRNA expression in rat testicular cell types. Molec. Cell. Endocrinol. 1992; 83:R1-R8.
58. Grootenhuis, A.J., Steenbergen, J., Timmerman, M.A., Dorsman, A.N.R.D., Schaaper, W.M.M., Meloen, R.H. and deJong, F.H. Activin. J. Endocrinol. 1989; 122:293-301.
59. Woodruff, T.K., Borrre, J., Attie, K.M., Cox, E.T., Rice, G.C. and Mather, J.P. Stage-specific binding of inhibin and activin to subpopulations of rat germ cells. Endocrinol. 1992;130:871-881.
60. Matthews, L.S. and Vale, W.W. Expression cloning of an activin receptor, a predicted transmembrane serine kinase. Cell 1991; 65:973-982.
61. deWinter, J.P., Themmen, A.P.N., Hoogerbrugge, J.W., Klaij, I.A., Grootegoed, J.A. and deJong, F.H. Activin receptor mRNA expression in rat testicular cell types. Mol. Cell Endocrinol. 1992; 83:R1-R8.
62. deWinter, J.P., Themmen, A.P.N., Hoogerbrugge, J.W., Klaij, I.A., Grootegoed, J.A. and deJong, F.H. Activin receptor mRNA expression in rat testicular cell types. Mol. Cell Endocrinol. 1992; 83:R1-R8.
63. Kaipia, A., Parvinen, M. and Toppari, J. Localization of activin receptor ACTR-IIB-2 mRNA in the rat seminiferous epithelium. Endocrinol. 1993; 132:477-479.
64. Ogawa, K., Kurohmaru, M., Shiota, K., Takahashi, M. and Hayashi, Y. Immunoreactive βA subunit of inhibin/activin is present in cytoplasm of rat spermatogenic cells. J. Reprod. Devel. 1992; 38:5-9.
65. Nakamura, T., Takio, K., Eto, Y., Shibai, H., Titani, K. and Sugino, H. Activin binding protein from rat ovary is follistatin. Science 1990; 247:836-838.

66. Lombardo, M., Hakky, S.I., Hall, M.K. and Hudson, P.B. A study of androgen biosynthesis by the human testis in vitro. J. Steroid Biochem. Molec. Biol. 1993; 44:191-198.
67. Luu-The, V., Labrie, C., Simard, J., Lachance, Y., Zhao, H.F., Conet, J., Leblanc, G. and Labrie, F. Structure of two in tandem human 17β-hydroxysteroid dehydrogenase genes. Molec. Endocrinol. 1990; 4:268-275.
68. Wright, W.W., Chan, K., Sundaram, K. and Bardin, C.W. New observations on androgen action: Androgen receptor stabilization and antisteroid effects of LHRH agonists. In: Leavitt, W.W. ed. Hormones and Cancer. NY: Plenum, 1982:325-336.
69. Kerr, J.B., Savage, G.N., Millar, M. and Sharpe, R.M. Response of the seminiferous epithelium of the rat testis to withdrawal of androgen: evidence for direct effect upon intercellular spaces associated with Sertoli cell junctional complexes. Cell Tiss. Res. 1993; 274:153-161.
70. Lyon, M.F., Glenister, P.H. and Lamoreux, M.L. Normal spermatozoa from androgen-resistant germ cells of chimeric mice and the role of androgen in spermatogenesis. Nature 1975; 258:620-622.
71. Tindall, D.J., Miller, D.A. and Means, A.R. Characterization of androgen receptors in Sertoli cell enriched testis. Endocrinol. 1977; 101:13-23.
72. Sanborn, B.M., Steinberger, A., Tcholakian, R.K., Tsai, Y.H. and Steinberger, E. Direct measurements of androgen receptors in cultured Sertoli cells. Steroids 1977; 29:493-502.
73. Anthony, C.T., Kovacs, W.J. and Skinner, M.K. Analysis of the androgen recptor in isolated testicular cell types with a microassay that uses an affinity ligand. Endocrinol. 1989; 125:2628-2635.
74. Naklha, A.M., Mather, J.P., Janne, O.A. and Bardin, C.W. Estrogen and androgen receptors in Sertoli, Leydig, myoid and epithelial cells: Effects of time in culture and cell density. Endocrinol. 1984; 115:121-128.
75. Namiki, M., Yokokawa, K., Okuyama, A., Koh, E., Kioyhara, H., Nakao, M., Sakoda, S. Matsumoto, K. and Sonoda, T. Evidence for the presence of androgen receptors in human Leydig cells. J. Steroid Biochem. Mol. Biol. 1991; 38:79-82.
76. Chang, C., Kokontis, J. and Liao, S. Structural analysis of complementary DNA and amino acid sequences of human and rat androgen receptors. Proc. Natl. Acad. Sci. USA 1988; 85:7211-7215.
77. Lubahn, D.B., Joseph, D.R., Sar, M., Tan, J., Higgs, H.N., Larson, R.E., French, F.S. and Wilson, E.M. The human androgen receptor: complementary deoxyribonucleic acid cloning, sequence analysis and gene expression in the prostate. Mol. Endocrinol. 1988; 2:1265-1275.
78. Jenster, G., van der Korput, H.A.G.M., van Vroonhoven, C., van der Kwast, T.H., Trapman, J. and Brinkman, A.O. Domains of the human androgen receptor involved in steroid binding, transcriptional activation, and subcellular localization. Mol. Endocrinol. 1991; 5:1396-1404.
79. Rundlett, S.E., Wu, W.-P. and Miesfield, R.L. Functional characterization of the androgen receptor confirms that the molecular basis of androgen action is transcriptional activation. Mol. Endocrinol. 1990; 4:708-714.
80. Lubahn, D.B., Brown, T.R., Simental, J.A., Higgs, H.N., Migeon, C.J., Wilson, E.M. and French, F.S. Sequence of the intron/exon junctions of the coding region of the human androgen receptor gene and the identification of a point mutation in a family with complete androgen insensitiv-

ity. Proc. Natl. Acad. Sci. USA 1989; 86:9534-9538.

81. Mowszowicz, I., Lee, H.-J., Chen, H.-T., Mestayer, C., Portois, M.-C., Cabrol, S., Mauvais-Jarvis, P. and Chang, C. A point mutation in the second zinc finger of the DNA-binding domain of the androgen receptor gene causes complete androgen insensitivity in two siblings with receptor-positive androgen resistance. Molec. Endocrinol. 1993; 7:861-869.
82. Nakao, R., Yanase, T., Sakai, Y., Haji, M. and Nawata, H. A single amino acid substitution Gly743->Val in the steroid-binding domain of the human androgen receptor leads to Reifenstein syndrome. J. Clin. Endocrinol. Metab. 1993; 77:103-107.
83. Nakao, R., Haji, M. and Yanase, T. A single amino acid substitution Met 786->Val in the steroid binding domain of human androgen receptor leads to a complete androgen insensitivity syndrome. J. Clin. Endocrinol. Metab. 1992; 74:1152-1157.
84. Brown, T.R., Lubahn, D.B., Wilson, E.M., Joseph, D.R., French, F.S. and Migeon, C.J. Deletion of the steroid-binding domain of the human androgen receptor gene in one family with complete androgen insensitivity syndrome: evidence for further genetic heterogeneity in this syndrome. Proc. Natl. Acad. Sci. USA 1988; 85:8151-8155.
85. Brown, T.R., Lubahn, D.B., Wilson, E.M., French, F.S., Migeon, C.J. and Corden, J.L. Functional characterization of naturally ocurring mutant androgen receptors from subjects with complete androgen insensitivity syndrome. Mol. Endocrinol. 1990; 4:1759-1772.
86. Marcelli, M., Tilley, W.D., Wilson, C.M., Griffen, G.E., Wilson, J.D. and McPhaul, M.J. A single nucleotide substitution introduces a premature termination codon into the androgen receptor gene of a patient with receptor negative androgen resistance. J. Clin. Invest. 1990; 85:1522-1528.
87. Marcelli, M., Tilley, W.D., Wilson, C.D., Griffin, J.E., Wilson, J.D. and McPhaul, M.J. Definition of the human androgen receptor gene structure permits the identification of mutations that cause androgen resistance: premature termination of the receptor protein at amino acid residue 588 causes complete androgen resistance. Mol. Endocrinol. 1990; 4:1105-1116.
88. Ris-Stalpers, C., Kuiper, G.G.J.M. and Faber, P.W. *et al.* Aberrant splicing of androgen receptor mRNA results in synthesis of a non-functional receptor protein in a patient with androgen insensitivity. Proc. Natl. Acad. Sci. USA 1990; 87:7866-7870.
89. Sai, T., Seino, S., Chang, C., Trifiro, M., Pinsky, L., Mhatre, A., Kaufman, M., Lambert, B., Trapman, J., Brinkmann, A.O., Rosenfield, R.L. and Liao, S. An exonic point mutation of the androgen receptor gene in a family with complete androgen insensitivity. Am. J. Hum. Genet. 1990; 46:1095-1100.
90. Marcelli, M., Tilley, W.D., Zoppi, S., Griffen, J.E., Wilson, J.D. and McPhaul, M.J. Androgen resistance associated with a mutation of the androgen receptor at amino acid 772 Arg->Cys results from a combination of decreased messenger ribonucleic acid levels and impairment of receptor function. J. Clin. Endocrinol. Metab. 1991; 73:318-325.
91. Marcelli, M., Zoppi, S., Grino, P.B., Griffin, J.E., Wilson, J.D. and McPhaul, M.J. A mutation in the DNA-binding domain of the androgen receptor gene causes complete testicular feminization in a patient with receptor-positive androgen resistance. J. Clin. Invest. 1991; 87:1123-1126.

92. Trifiro, M., Prior, R.L., Sabbaghian, N,, Pinsky, L., Kaufman, M., Nylen, E.G., Belsham, D.D., Greenberg, C.R. and Wrogemann, K. Amber mutation causes a diagnostic Mael site in the androgen receptor gene of a family with complete androgen insensitivity. Am. J. Med. Genet. 1991; 40:493-499.
93. Zoppi, S., Marcelli, M., Deslypere, J.-P., Griffen, J.E., Wilson, J.D. and McPhaul, M.J. Amino acid substitutions in the DNA-binding domain of the human androgen receptor are a frequent cause of receptor-binding positive androgen resistance. Mol. Endocrinol. 1992; 6:409-415.
94. Quidgley, C.A., Evans, B.A.J., Simental, J.A., Marschke, K.B., Sar. M., Lubahn, D.B., Davis, P., Hughes, I.A., Wilson, E.M. and French, F.S. Complete androgen insensitivity due to deletion of exon C of the androgen receptor gene highlights the functional importance of the second zinc finger of the androgen receptor in vitro. Mol. Endocrinol. 1992; 6:1103-1112.
95. Ris-Stalpers, C., Trifiro, M.A., Kuiper, G.G.J.M., Jenster, G., Romalo, G., Sai, T., van Rooij, H.C.J., Kaufman, M., Rosenfield, R.L., Liao, S., Schwekkert, H.-U., Trapman, J., Pinsky, L. and Brinkman, A.O. Substitution of aspartic acid-686 by histidine or asparagine in the human androgen receptor leads to a functionally inactive protein with hormone-binding characteristics. Mol. Endocrinol. 1991; 5:1562-1569.
96. Griswold, M.D., Bishop, K.-H., Kim, R., Ping, P.E., Sliteri, J.E. and Morales, C. Function of vitamin A in normal and synchronized seminiferous tubules. Ann. N.Y. Acad. Sci. 1989; 564: 154-172.
97. Kim, K.H. and Griswold, M.D. The regulation of retinoic acid receptor mRNA levels during spermatogenesis. Mol. Endocrinol. 1990; 4: 1679-1688.
98. van Pelt, A.M.M., van den Brink, C.E., de Rooij, D.G. and van der Saag, P.T. Changes in retinoic acid receptor messenger ribonucleic acid levels in the vitamin A-deficient rat testis after administration of retinoids. Endocinol. 1992; 131:344-350.
99. Lufkin, T., Lohnes, D., Mark, M., Dierich, A., Gorry, P., Gaub, M.-P., LeMeur, M. and Chambon, P. High postnatal lethality and testis degeneration in retinoic acid receptor α mutant mice. Proc. Natl. Acad. Sci. USA 1993; 90:7225-7229.
100. Huggenvik, J. and Griswold, M.D. Retinol-binding protein in rat testicular cells. J. Reprod. Fertil. 1981; 61:403-408.
101. Schmitt, M.C. and Ong, D.E. Expression of cellular retinol-binding protein and lecithin-retinol acyltransferase in developing rat testis. Biol. Reprod. 1993; 49:972-979.
102. Davis, J.T. and Ong, D.E. Synthesis and secretion of retinol-binding protein by cultured rat Sertoli cells. Biol. Reprod. 1992; 47:528-533.
103. Li, S.S.L., O'Brien, D.A., Hou, E.W., Versola, J., Rockett, D.L. and Eddy, M. Differential synthesis of lactate dehydrogenase isozymes A (muscle), B (heart) and C (testis) in mouse spermatogenic cells. Biol. Reprod. 1989;44: 173-180.
104. Adham, I.M., Burkhardt, E., Benahmed, M. and Engel, W. Cloning of a cDNA for a novel insulin-like peptide of the testicular Leydig cells. J. Biol. Chem. 1993; 268:26668-26672.
105. Bellve, A.R. and Zheng, W. Growth factors as autocrine and paracrine

modulators of male gonadal functions. J. Reprod. Fertil. 1989; 85:771-793.

106. Robertson, D.M., Risbridger, G.P., Hedger, M. and McLachlan, R.I. Growth factors in the control of testicular function. In: deKretser, D. ed. Molecular Biology of the Male Reproductive System. NY: Acad.Press, 1993:411-438.

107. Mullaney, B.P. and Skinner, M.K. Transforming growth factor-α and epidermal growth factor receptor gene expression and action during pubertal development of the seminiferous tubule. Molec. Endcrinol. 1992; 6:2103-2113.

108. Skinner, M.K., Takacs, K. and Coffey, R.J. Cellular localization of transforming growth factor-α gene expression and action in the seminferous tubule: Peritubular cell-Sertoli cell interactions. Endocrinol. 1989; 124:845-854.

109. Teerds, K.J., Rommerts, F.F.G. and Dorrington, J.H. Immunohistochemical detection of transforming growth factor-α in Leydig cells during the development of the rat testis. Mol. Cell. Endocrinol. 1990; 69:R1-R6.

110. Wong, S.T., Winchell, L.F., McCune, B.K., Earp, S.H., Teixido, J., Massague, J., Herman, B. and Lee, D.C. The TGF-α precursor expressed on the cell surface binds to the EGF receptor on adjacent cells, leading to signal transduction. Cell 1989; 56:495-506.

111. Mullaney, B.P. and Skinner, M.K. Transforming growth factor-β b1, b2, and b3 gene expression and action during pubertal development of the seminiferous tubule: Potential role at the onset of spermatogenesis. Molec. Endocrinol 1993; 7:67-76.

112. Teerds, K.J. and Dorrington, J.H. Localization of transforming growth factor β-1 and β-2 during testicular development in the rat. Biol. Reprod. 1993; 48:40-45.

113. Skinner, M.K. and Moses, H.L. Transforming growth factor-β gene expression and action in the seminiferous tubule: peritubular cell-Sertoli cell interactions. Mol. Endocrinol. 1989; 3:625-634.

114. Watrin, F., Scotto, L., Assoian, R.K. and Wolgemuth, D.J. Cell lineage specificity of expression of the murine transforming growth factor β3 and transforming growth factor β1 genes. Cell Growth Differ. 1991; 2:77-83.

115. Griswold, M.D. Protein secretion of Sertoli cells. Int. Rev. Cytol. 1988; 110:133-156.

116. Radhakrishnan, R., Oke, B.O., Papadopoulos, V., DeAugustine, R.P. and Suarez-Quian, C.A. Characterization of epidermal growth factor in mouse testis. Endocrinol. 1992; 131:3091-3099.

117. Tsutami, O., Kurachi, H. and Oka, T. A physiological role of epidermal growth factor in male reproductive function. Science 1986; 233:975-978.

118. Radhakrishnan, B. and Suarez-Quian, C.A. Characterization of epidermal growth factor receptor in testis, epididymis and vas deferens of non-human primates. J. Reprod. Fertil. 1992; 96:13-23.

119. Suarez-Quian, C.A. and Niklinski, W. Immunocytochemical localization of the epidermal growth factor receptor in mouse testis. Biol. Reprod. 1990; 43:1087-1097.

120. Suarez-Quian, C.A., Dai, M., Onoda, M., Kriss, R.M. and Dym, M. Epidermal growth factor receptor localization in the rat and monkey testis. Biol. Reprod. 1989; 41:921-932.

121. Lamb, D.J. Growth factors and testicular development. J. Urol. 1993;

150:583-592.

122. Damjanov, I., Solter, D. and Knowles, B.B. Functional epidermal growth factor receptor localizes to the postacrosomal region of human spermatozoa. Biochem. Biophys. Res. Comm. 1993; 190: 901-906.
123. Lahr, G., Mayerhofer, A., Seidl, K., Bucher, S., Grothe, C., Knochel, W. and Gratzl, M. Basic fibroblast growth factor bFGF in rodent testis. Presence of bFGF mRNA and of a 30 kDa bFGF protein in pachytene spermatocytes. FEBS Lett. 1992; 302:43-46.
124. Han, I.S., Sylvester, S.R., Kim, K.H., Schellng, M.E., Venkateswaran, S., Blanckaert, V.D., McGuiness, M.P. and Griswold, M.D. Basic fibroblast growth factor is a testicular germ cell product which may regulate Sertoli cell function. Molec. Endocrinol. 1993; 7:889-897.
125. Gnessi, L., Emidi, A., Farini, D., Scarpa, S., Modesti, A., Ciampani, T., Silvestroni, L. and Spera, G. Rat Leydig cells bind platelet-derived growth factor through specific receptors and produce platelet-derived growth factor-like molecules. Endocrinol. 1992; 130:2219-2224.
126. Gnessi, L., Emidi, A., Scarpa, S., Palleschi, S., Ragano-Caracciolo, M., Silvestroni, L., Modesti, A. and Spera, G. Platelet-derived growth factor effects on purified testicular myoid cells: Binding, cytosolic Ca^{2+} increase, mitogenic activity and extracellular matrix production enhancement. Endocrinol. 1993; 133:1880-1890.
127. Tres, L.L., Smith, E.P. and Kierszenbaum, A.L. Immunoreactive sites and accumulation of somatomedin-C in rat Sertoli-spermatogenic cell co-cultures. Exp. Cell Res. 1986; 162:33-50.
128. Moore, A., Chen, C.-L. C., Davis, J.R.E. and Morris, I.D. Insulin-like growth factor-I mRNA expression in the interstitial cells of the rat testis. J. Molec. Endocrinol. 1993; 11:319-324.
129. Smith, E.P., Svoboda, M.E., VanWyk, J.J., Kierszenbaum, A.L. and Tres, L.L. Partial characterization of a somatomedin-like peptide from the medium of cultured rat Sertoli cells. Endocrinol. 1987; 120:186-193.
130. Spiteri-Grech, J. and Nieschlag, E. The role of growth hormone and insulin-like growth factor I in the regulation of male reproductive function. Horm. Res. 1992; 38 Supp 1:22-27.
131. Zhou, J. and Bondy, C. Anatomy of the insulin-like growth factor system in the human testis. Fertil. Steril. 1993; 60:897-904.
132. Vanelli, B.G., Barni, T., Orlando, C., Natali, A., Serio, M. and Balboni, G.C. Insulin-like growth factor-I IGF-I and IGF-I receptor in human testis: An immunochemical study. Fertil. Steril. 1988; 49:666-669.
133. Ritzen, E.M. Chemical messengers between Sertoli cells and neighbouring cells. J. Steroid Biochem. 1983; 19:499-504.
134. Tres, L.L., Smith, E.P., van Wyk, J.J. and Kierszenbaum, A.L. Immunoreactive sites and accumulation of somatomedin-C in rat Sertoli-spermatogenic cell co-cultures. Exp. Cell Res. 1986; 162:33-50.
135. Moore, A. and Morris, I.D. The involvement of insulin-like growth factor-I in local control of steroidogenesis and DNA synthesis of Leydig and non-Leydig cells in the rat testicular interstitium. J. Endocrinol. 1993; 138:107-114.
136. Smith, E.P., Dickson, B.A. and Chernausek, S.D. Insulin-like growth factor binding protein-3 secretion from cultured rat Sertoli cells: dual regulation by follicle stimulating hormone and insulin-like growth factor-I.

Endocrinol. 1990; 127:2744-2755.

137. Skalli, M., Avallet, O., Vigier, M. and Saez, J.M. Opposite vectorial secretion of insulin-like growth factor I and its binding proteins by pig Sertoli cells cultured in the bicameral chamber system. Endocrinol. 1992; 131:985-987.
138. Choongkittaworn, N.M., Kim, K.H. Danner, D.B. and Griswold, M.D. Expression of prohibitin in rat seminiferous epthelium. Biol. Reprod. 1993; 49:300-310.
139. Pollanen, P., Soder, O. and Parvinen, M. Interleukin-1 α stimulation of spermatogonial proliferation in vivo. Reprod. Fertil. 1989; 1:85-87.
140. Soder, O., Syed, V., Callard, J., Toppari, J., Pollanen, P., Parvinen, M., Froysa, B. and Ritzen, E.M. Production and secretion of an interleukin-1-like factor. Int. J. Androl. 1991; 14:223-231.
141. Parvinen, M., Soder, O., Mali, P., Froysa, B. and Ritzen, E.M. In vitro stimulation of stage-specific DNA synthesis in rat seminiferous tubule segments by interleukin-1α. Endocrinol. 1991; 129:1614-1620.
142. Syed, V., Gerard, N., Kaipia, A., Bardin, C.W., Parvinen, M. and Jegou, B. Identification, ontogeny, and regulation of an interleukin-6-like IL-6 factor in the rat testis. Endocrinol. 1993; 132:293-299.

142a.Takao, T., Mitchell, W.M., W.M., Tracey, D.E. and de Souza, E.B. Identification of interleukin-1 receptors in mouse testis. Endocrinol. 1990; 127:251-258.

143. Parvinen, M., Pelto-Huikko, M., Soder, O., Schultz, R., Kaipia, A., Mali, P., Toppari, J., Hakovirta, H., Lonnerberg, P., Ritzen, W.M., Ebendal, T., Olson, L., Hokfelt, T. and Persson, H. Expression of β-nerve growth factor and its receptor in rat seminiferous epithelium: Specific function at the onset of meiosis. J. Cell Biol. 1992; 117:629-641.
144. Persson, H., Ayer-LeLievre, C., Soder, O., Villar, M.J., Metsis, M., Olson, L., Ritzen, E.M. and Hokfelt, T. Expression of β-nerve growth factor receptor mRNA in Sertoli cells downregulated by testosterone. Science 1990; 247:704-707.
145. Hempstead, B.L., Martin-Zanca, D., Kaplan, D.R., Parada, L.F. and Chao, M.V. High affinity NGF binding requires coexpresion of the *trk* proto-oncogene and the low-affinity NGF receptor. Nature 1991; 350:678-683.
146. Ayer-LeLievre, C., Olson, L., Ebenda, T., Hallbood, F. and Persson, H. Nerve growth factor mRNA and protein in the testis and epididymis of mouse and rat. Proc. Natl. Acad. Sci. USA 1988; 85:2628-2632.
147. Verhoeven, G. and Cailleau, J.A. A factor in spent media from Sertoli cell-enriched cultures that stimulates steroidogenesis in Leydig cells. Mol. Cell. Endocrinol. 1985; 40:57-68.
148. Feig, L.A., Bellve, A.R., Horback-Erickson, N. and Klagsbrun, M. Sertoli cells contain a mitogenic polypeptide. Proc. Natl. Acad. Sci. USA 1980; 77:4774-4778.
149. Feig, L.A., Klagsbrun, M. and Bellve, A.R. Mitogenic polypeptide of the mammalian seminiferous epithelium: biochemical characterization and partial purification. J. Cell Biol. 1983; 97:1435-1440.

CHAPTER 6

Regulation of Spermatogenesis

INTRODUCTION

Regulation of the orderly progression of male germ cell development involves contributions from endocrine somatic cells of the pituitary, testicular somatic cells of the interstitium and seminiferous epithelium, and the germ cells themselves. There are many possibilities for developmental and physiological controls, including positively and negatively acting elements and feedback loops. Critical decisions begin with sex determination, the choice to develop a testis instead of an ovary. For production of large numbers of relatively short-lived spermatozoa, stem cells must be kept in reserve and recruited as necessary. Levels of hormones and growth factors must be regulated. In this chapter, mammalian sex determination, control of stem cell proliferation, and interactions of the pituitary and testis, of the testicular somatic cells, and of the germ cells and testicular somatic cells is considered with reference to potential regulatory mechanisms. For a review of regulation of invertebrate spermatogenesis, see ref. 1.

SEX DETERMINATION

In mammals, the Y chromosome is required and sufficient for male development no matter how many X chromosomes are present. The gene (or genes) on the Y chromosome responsible for male sex determination is the called *testis-determining factor:* TDF in humans or Tdy in mice. TDF is believed to cause supportive somatic cells of the indifferent gonad to differentiate as Sertoli cells.[2] For reviews of mammalian sex determination, see refs. 3 and 4.

The gene *SRY* (sex-determining region Y-linked gene) maps to the smallest part of the Y chromosome which is sex-determining.[5,6] The open reading frame codes as a single exon for a 204-amino acid protein in humans, 395 amino acids in mice. Homology in a central 79-amino acid region exists between human, mouse and rabbit. This region encodes an HMG box shared with the non-histone nuclear protein HMG1 and several other proteins, and constitutes a DNA-binding domain. SRY recognizes the sequences AACAAAG and TTCAAAG in double stranded DNA.[7,8] The regions outside the HMG box are poorly conserved and indeed have evolved rapidly.[9,10] The most likely role for SRY is as a DNA binding factor affecting transcription of downstream genes. Whether SRY acts as an activator or repressor of transcription, or even in which cell types it is expressed, remain to be elucidated.

SRY-related genes have been isolated from testis using degenerate primers and the polymerase chain reaction.[11] *Sox-5* encodes a putative transcription factor of 43 kDa containing an HMG-box. The protein is expressed at high levels in round spermatids and is not expressed premeiotically. It can bind to the consensus sequence also recognized by SRY.

STEM CELL PROLIFERATION

A common technique for synchronization of mammalian spermatogenesis is removal of vitamin A from the diet.[12] After loss of advanced germ cells, spermatogenesis is reinitiated from Type A1 spermatogonia upon addition of retinol.[13] Four hrs after retinol treatment of rats, cdc2/kinase/cyclin B activity as a mitotic indicator increases reaching a maximum at 12 hr,[14] suggesting the cells are blocked at the end of S phase and traverse G2 in 11 hr.

Two genes involved in control of stem cell proliferation in mouse spermatogenesis are the *white spotting locus* (*W*) and the *Steel locus* (*Sl*).[15-17] Loss of function mutations in these genes affect three developmental processes: hematopoiesis, melanogenesis and gametogenesis. The *W* gene encodes *c-kit*, a tyrosine kinase growth factor related to PDGF and homologue of the viral oncogene *v-kit*. The *Sl* gene encodes the kit ligand (KL) or Steel factor. It is expressed in somatic cells along the route followed by the germ cells during migration to the gonad and also in the genital ridge.[18]

Injection into the mouse peritoneal cavity of antibody to *c-kit* blocks binding to the KL and inhibits Type A spermatogonial proliferation and differentiation.[19] *c-Kit* is expressed in primordial germ cells during migration, proliferation and differentiation phases.[20] *c-Kit* is expressed at high levels in Type A2 and Type B spermatogonia prior to entry into meiosis, and is also expressed in Leydig cells.[21,22] In mouse testis, *c-kit* is encoded by a 5.5 kb mRNA from spermatogonia to pachytene spermatocytes.[21] It encodes a 145 kDa protein whose developmental expression parallels that of the mRNA.

KL is necessary for primordial germ cell viability in vitro and promotes PGC proliferation.[23] KL exists in both a soluble and a membrane-associated form encoded by distinct mRNAs.[24] The mRNA for the membrane bound form of KL is more abundant in 6 day mouse Sertoli cells than the soluble form.

One mode of action of the two gene products is believed to be through a triggering of *c-kit* by KL binding to its extracellular domain, followed by autophosphorylation of the *c-kit* cytoplasmic tyrosines, followed by a signal transduction pathway involving phospholipase C-γ1.[16,17] A second mode may involve direct binding. Sertoli cells of Sl/Sl^d mutants cannot bind to germ cells and are deficient in the mRNA for the membrane form of KL. A microinjected plasmid containing the gene for the membrane form but not the soluble form can rescue binding in the mutant cells.[25]

mRNAs of 3.2 kb and 2.3 kb encode truncated forms of *c-kit* in spermatids[26] and may play a role in spermiogenesis.[27,28] Chimeric mice with normal and mutant Sl cells exhibit degeneration of various late stage cells suggesting a function for *c-kit* and KL in maintenance of spermiogenesis.

ENTRY INTO MEIOSIS

Little is known about control of meiosis by local factors in the testis. Sertoli cells may produce a meiosis-preventing substance,[29] assayed by its ability to prevent meiosis in oocyte organ culture. It resembles steroids and may act through an inositol signaling pathway.[30] Mouse ovary contains substances that induce meiosis in male cells.[31]

PITUITARY-TESTIS INTERACTIONS

Pituitary hormones are needed for initiation and maintenance of spermatogenesis. Inhibins, activins and follistatin from the testis control pituitary function (Fig. 5.4). The latter have other more local effects in the testis involving somatic cell-germ cell interactions as well.

EFFECTS OF GONADOTROPINS ON TESTIS

The two major gonadotropins which affect testicular function are follicle-stimulating hormone and luteinizing hormone. Effects which have been studied are the initiation of spermatogenesis in immature animals, the maintenance of spermatogenesis in adults, and the reinitiation of spermatogenesis following hypophysectomy or treatment with antagonists of gonadotropin releasing hormones. The major testicular target cells for FSH are the Sertoli cells and for LH the Leydig cells. Hypophysectomy causes regression of testes in adult animals and inhibits the initiation of spermatogenesis in immature organisms. These effects can sometimes be reversed by FSH and LH. Results may be dependent on the organism studied.

Follicle-Stimulating Hormone

FSH can initiate spermatogenesis in immature rat testis. The ratio of spermatocytes to spermatids assayed by flow cytometry was reduced by 90% following immunization of rats against FSH.[32] FSH affects somatic cells of the immature testis, stimulating proliferation of Sertoli cells.[33] There is contradictory evidence concerning the role of FSH in maintaining adult spermatogenesis. Some studies indicate that immunization against FSH did not alter spermatogenesis[34] and restoration of FSH to hypophysectomized rats did not rescue spermatogenesis,[35] but others report FSH supported spermatogenesis up to spermatid Stage VII.[36] Maintenance of spermatogenesis and reinitiation after administration of GnRH antagonists was shown in rhesus monkeys.[37] FSH was not required to maintain spermatogenesis in the adult rat, but may act synergistically with testosterone.[38]

FSH causes major alterations in Sertoli cell function. FSH decreases Sertoli TGF-β2 mRNA without affecting TGF-β1 or TGF-β3.[39] Sertoli cells are stimulated by FSH to make IGF-I.[40] FSH does not regulate IGF-I receptors on Sertoli cells.[40] FSH transiently down regulates androgen receptor mRNA but not the protein in immature rats. This is followed by long term up-regulation of both receptor mRNA and protein.[41] FSH stimulates SCSGF secretion by Sertoli cells.[42] (SCSGF is also stimulated by testosterone, retinol and insulin.[43,44]) FSH and cAMP increase KL or Steel factor mRNA (both soluble and membrane bound forms) in cultured Sertoli cells.[45] FSH stimulates inhibin-α but not -β_B mRNA in vivo and in cultured Sertoli cells.[46] Inhibin-α but not -β_B mRNA increased in human testicular cells in response to gonadotropin preparations containing predominantly FSH or LH in vivo and in vitro.[47]

FSH stimulation of Sertoli cells to produce cAMP is inhibited by adenosine and analogs in a stage-specific fashion suggesting a possible paracrine role for germ cells.[48] The adenosine effect is mediated by the Gi subunit of a G-protein complex.[48]

FSH can affect germ line function through cAMP mediated signaling pathways as well. FSH but not LH or prolactin causes a switch in the cyclic-AMP-responsive element modulator (CREM) gene expression by alternative splicing in the male germ line. The gene product, a regulator of cAMP-transcriptional responses, is converted from an antagonist to an activator beginning in premeiotic spermatocytes.[49] The activator CREMτ has two inserted glutamine-rich domains which confer transcriptional activation activity.[50] Accumulation of high levels of messenger is caused by alternative polyadenylation using the most 5' polyadenylation site and enhances the stability of the mRNA.[49]

FSH also affects peritubular cells. Using α smooth muscle actin as a peritubular marker, the development of peritubular cells in monkeys was shown to respond to FSH and testosterone.[51] Maintenance of differentiation was not dependent on further exposure to the hormones.[51]

Luteinizing Hormone

LH is the major regulator of Leydig cell functions which include steroidogenesis. LH receptors, and the pathways to produce androgens, are acquired by immature Leydig cells as they start to differentiate[52] and these events are probably independent of LH.[53] The triggers of Leydig cell differentiation are not known.[54]

LH can restore Leydig cell inhibin-α mRNA in hypophysectomized rats.[47] LH regulates both IGF-I mRNA and protein synthesis and IGF-I receptors in and on Leydig cells.[55-57] IGF-I in turn stimulates cultured immature Leydig cells to produce LH binding activity and to synthesize testosterone.[58] LH receptors can be down-regulated after prolonged LH stimulation. Coupling to the Gs subunit of adenyl cyclase is lost.[59] Desensitization is seen only in adult but not fetal Leydig cells.[52]

Other hormones with effects on Leydig cells include prolactin, growth hormone, insulin, vasopressin, oxytocin, atrial natriuretic peptide ANP, various steroids and small Leydig cell inhibitors (for review, see ref 54). Leydig cells have receptors for prolactin,[60] vasopressin and oxytocin,[61] insulin,[62] and ANP.[63] Growth hormone receptors have been demonstrated in rat testis.[64]

Gonadotropins and Programmed Cell Death

Spermatogenesis normally includes programmed cell death of spermatogonia and meiotic cells.[65] Cell death occurs both in the seminiferous tubules and interstitial cells. Hypophysectomy causes cell death in vivo but cells may be rescued by FSH or hCG. The GnRH antagonist azaline B also increases apoptosis. FSH and LH/hCG regulate cell survival in the testis by preventing apoptosis.[65]

In the rat, GnRH antagonists resulted in degeneration of preleptotene and pachytene spermatocytes and Steps VII and XIX spermatids initially.[66] Eventually the loss of all advanced spermatids mimics the effect of hypophysectomy. A number of compounds which mimic the amino acid sequence of GnRH (Fig. 5.1) have been synthesized which are agonists or antagonists. GnRH stimulates the synthesis and release of FSH and LH. GnRH agonists stimulate gonadotropins and testosterone production initially, but subsequently testosterone levels decline. The antagonists on the other hand act immediately to lower gonadotropin and testosterone levels.[67] The results in either case are complete blockade of spermatogenesis and the induction of azoospermia in primates.

Testosterone can help prevent the effects of hypophysectomy. Hypophysectomy followed by immediate replacement with testosterone maintains Sertoli-spermatid junctional complexes, but if delayed it is not able to rescue, suggesting that testosterone maintains binding competence but cannot restore it.[68] High doses of GnRH antagonists impaired final maturation of spermatids in rats which was prevented by exogenous testosterone. The treatment restored ABP activity in interstitial fluid suggesting ABP availability is important for completion of maturation.[69]

Effects of Testis on Pituitary Function

The synthesis and release of gonadotropins is feedback regulated by inhibins, activins and follistatin (Fig. 5.4). The occurrence of extragonadal sources of these proteins however suggests caution in interpretation of the origin of the feedback loop.

Just as gonadotropins from the pituitary regulate the testis, the gonadal proteins inhibin, activin and follistatin from the testis feed-back regulate the pituitary. Inhibin secreted by Sertoli cells binds in the pituitary and suppresses FSH but not LH synthesis and release.[70,71] For further reviews, see refs. 47, 72-74. Activins stimulate the synthesis and release of FSH.[47,75] They are secreted by Sertoli and many other cell types.[76] Follistatins, biochemically unrelated to inhibin or activin, inhibit the release of FSH by cultured pituitary cells and modulate activin activity.[77]

Inhibin decreases FSHβ mRNA in the pituitary.[78] In cultured pituitary cells inhibin decreases and activin increases FSHβ mRNA.[79] Depletion of inhibin by immunoneutralization in immature rats increases FSHβ mRNA.[80] In addition to effects on FSH, activins can also inhibit growth hormone biosynthesis[81] and secretion of growth hormone releasing hormone,[82,83] ACTH[84] and prolactin[85] by the pituitary.

SOMATIC CELL-SOMATIC CELL INTERACTIONS

There are many possible routes of autocrine and paracrine interactions among somatic cells of the testis. However, there are a number of complicating factors to be considered in evaluating interactions. Many potential regulators are produced by more than one cell type, making it difficult or impossible to establish significant pathways. The presence of molecules in excess may be permissive for a given biological function, but not involve a regulatory step. Furthermore, levels of products may change with stage of maturity of the gonad. Finally, purified cell types under investigation may not behave normally in isolation, and they are sometimes contaminated with other cell types. Possible routes of interaction among somatic testicular cells are discussed below (Fig. 6.1).

DIFFUSIBLE FACTORS

Several diffusible factors primarily affect differentiation, such as androgens, PModS, and various POMC peptides. Others affect growth or proliferation such as IGF, TGF-α, TGF-β, NGF, FGF, and SGF. However, even this division may be oversimplified.

The cell types which have received virtually all the attention are the Sertoli, Leydig and peritubular cells, but other cells, such as macrophages, may also play important roles.[86] In some cases the sources of diffusible factors remains uncharacterized. For example, unknown factors present in medium from androgen treated cultured cells from testis inhibit steroid production by neonatal Leydig cells.[87] In other cases, there are documented multiple sources, such as for TGF and IGF made by Sertoli, Leydig and peritubular cells. Potential somatic cell interactions through diffusible substances in the testis have been recently reviewed.[88]

Sertoli-Peritubular Interactions

Both Sertoli and peritubular cells produce TGF-α and TGF-β.[89-91] EGF receptors (which bind TFG-α) have been located on peritubular cells and may be present on Sertoli cells as well.[89,92] TGF-α/EGF stimulates peritubular cells, but not mid-pubertal Sertoli cells.[89] TGF-α increases a specific plasminogen activator inhibitor (PAI-1) mRNA in peritubular cells which is presumed to encode a molecule which breaks down basement lamina, but it decreases tissue type plasminogen activator mRNA in Sertoli cells.[93,94] Overexpression of TGF-α in transgenic mice results in no obvious abnormalities in testis structure or spermatogenesis, however.[86] TGF-β has little or no effect on Sertoli cells, but modulates TGF-α stimulation of peritubular cell proliferation and differentiation.[39,90,93] TGF-β also affects peritubular cell migration, colony formation and matrix production in rats.[90,95]

IGF-I is also made by both Sertoli and peritubular cells and stimulates the growth of peritubular and immature Sertoli cells.[88] IGF-I stimulates the formation of plasminogen activator, but inhibits transferrin production in Sertoli cells.[96,97] Peritubular cells make IGF binding protein-2 and Sertoli cells make IGF binding protein-[398], carriers of IGF in serum and extracellular fluids. Interference with normal Sertoli function by lowering the numbers of late spermatids alters peritubular cell shape and organization.[99]

Peritubular cells, but not germ cells, enhance Sertoli cell attachment and secretion of SCSGF.[42] PModS from peritubular cells stimulates several Sertoli functions. It inhibits induction of aromatization activity,[100] stimulates secretion of APB,[89] and results in increased Sertoli cell c-*fos* expression.[101]

Leydig-Sertoli Interactions

Sertoli cells can affect Leydig cell development, proliferation and patterns of gene expression. Leydig cells can modulate Sertoli cells through testosterone or POMC peptides. Some effects, studied in vitro with isolated Sertoli and Leydig cell cocultures must be regarded with some caution since removing germ cells, which interact with Sertoli cells and metabolize testosterone, can make the system behave abnormally. For reviews, see refs. 102-104.

Sertoli cells affect Leydig cells in several ways. Increases in Leydig cell numbers and their maturation during early puberty probably depend on growth factors from Sertoli cells. One possible factor is a 30 kDa protein which stimulates Leydig cell DNA synthesis.[105] Loss of germ cells may affect Leydig cells through the mediation of Sertoli cells. Disruption of spermatogenesis by a variety of means results in hypertrophy of Leydig cells in the damaged regions, regeneration of Leydig cells after chemical treatment is accelerated in the absence of germ cells, and experimentally induced increases in neonatal Sertoli cell numbers result in increased numbers of Leydig cells (reviewed in ref. 106).

]Leydig and Sertoli cells both make estrogens, inhibins, activin, IGF-I, TGF-α and TGF-β, so paracrine and autocrine effects are not clear.[88] Sertoli cells produce estradiol during fetal and neonatal life and Leydig cells arc thc main source subsequently.[107] Estradiol prevents regeneration of Leydig cells in adult rats by suppressing precursor cell proliferation, suggesting an autocrine regulation.[108] IGF-I, inhibin and TGF-α all stimulate steroidogenesis in Leydig cells.[109] In another potential autocrine effect, IGF-I increases LH binding sites on Leydig cells while down-regulating its own receptors.[56,96,110] IGF-I also enhances androgen production by immature Leydig cells which are more sensitive than adult Leydig cells.[109]

Sertoli cells produce IL-1α and Leydig cells have IL-receptors.[110a] EGF affects both Sertoli and Leydig cells. Sertoli cells have been reported to possess or lack EGF receptors.[92,111] EGF may alter lactate and inhibin production in Sertoli cells[112,113] but the effect may be indirect.[114]

IL-1 and LHRH-like activities produced by Sertoli cells affect Leydig cells by, for example, inhibiting testosterone secretion.[115,116] In general, coculture of Sertoli with Leydig cells results in increased LH triggered testosterone production. Leydig cell stimulatory factor from cultured Sertoli cell medium[117] stimulates steroidogenesis in Leydig cells. Sertoli cells secrete Leydig cell stimulator which stimulates testosterone production[118] and Leydig cell inhibitor which depresses cAMP levels and hCG stimulated androgen production of Leydig cells.[118] IGF-I produced by Sertoli or peritubular cells, acting synergistically with TGF-α and insulin, binds to Leydig cells and increases steroidogenesis and DNA synthesis.[119]

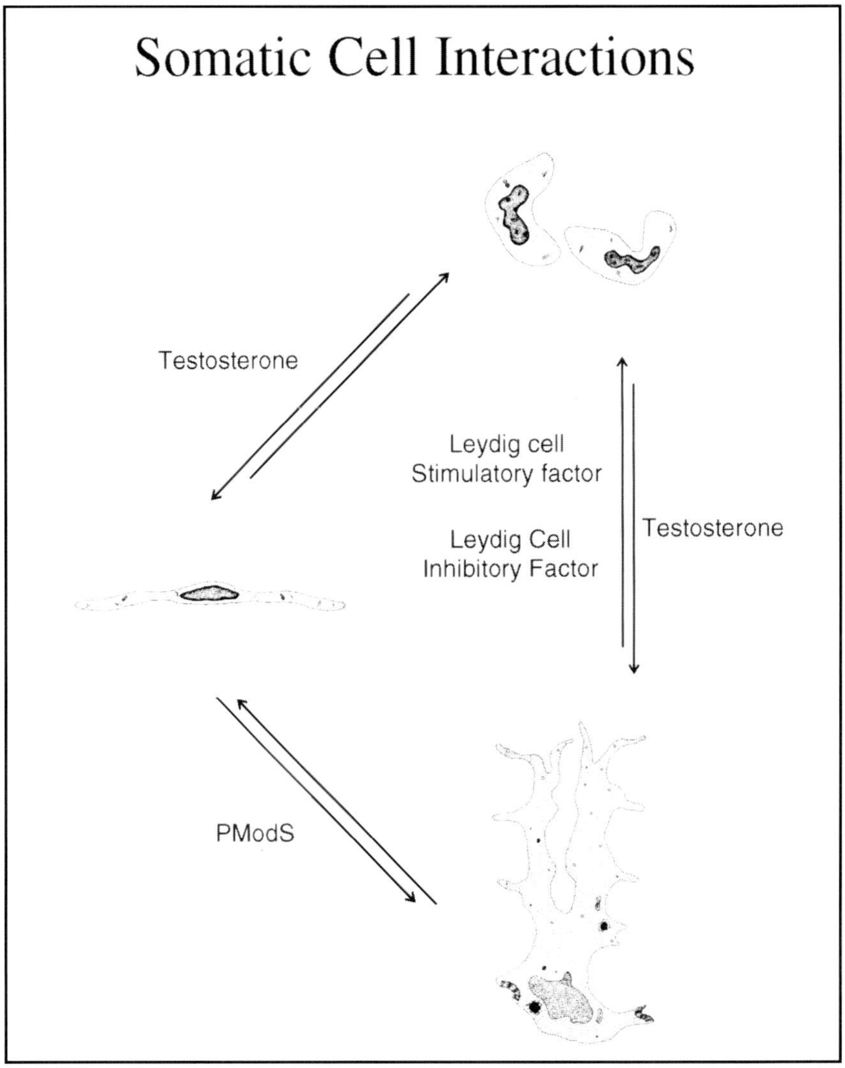

Fig. 6.1. Some interactions of the somatic cells of the testis. Many other diffusible factors are likely to mediate interactions of the somatic cells, but their originating cell types are less clear. (Drawing by J. Poccia.)

The *c-kit* ligand KL is produced in Sertoli cells.[120] Sertoli cell expression is higher in germ cell-deficient gonads suggesting negative regulation by germ cells. In Leydig cells a 5.5 kb *c-kit* mRNA is found.[21,22,27] When *c-kit*-KL interactions are blocked by injection of anti-KL, testosterone in the serum increases implying a role in suppressing Leydig cell function.[19]

Factors in interstitial fluid presumed to be produced by Sertoli cells stimulate steroidogenesis in Leydig cells.[121,122] Prepubertal Sertoli cell-secreted factors stimulate proliferation in cultured Leydig cells and inhibit T and B lymphocyte proliferation.[123] Sertoli cells in adult rats secrete an aromatase inhibitor which can affect Leydig cell proliferation.[124]

The transcription factor *c-jun* is expressed in cultured pig Leydig cells and is stimulated by bFGF and EGF. *JunB* expression is stimulated by bFGF, EGF, hCG and IGF-I.[125] *junD* is also expressed in Leydig cells of mouse.[126]

Leydig cell testosterone is a major regulator of Sertoli cells. Sertoli cells in culture are stimulated by testosterone to elevate RNA polymerase II levels and to synthesize a variety of proteins.[127,128] Testosterone also decreases plasminogen activator secretion and ornithine decarboxylase mRNA levels in immature cultured Sertoli cells.[129,130] Testosterone and other factors stimulate Sertoli cell ABP activity[131] but the effects may be indirect. Chronic testosterone effects on adult testis suggest stimulation of Sertoli transferrin mRNA is indirect, mediated by increasing germ cell numbers.[38] The gene for low-affinity NGF-β receptors in Sertoli cells is down-regulated by testosterone.[132] Testosterone stimulates secretion of Sertoli testins in vitro.[133] Testosterone also stimulates secretion of Sertoli cell secreted growth factor (SCSGF).[42] Withdrawal of testosterone by hypophysectomy or chemical killing of Leydig cells results in alteration of Sertoli junctional complexes and appearance of fluid filled spaces between Sertoli cells.[134]

Leydig cells may affect Sertoli cells through production of testosterone or proopiomelanocortin (POMC) peptides. POMC peptides including β-endorphin, α-melanocyte stimulating hormone and adrenocorticotropin (ACTH) are produced by Leydig cells.[135] Sertoli cells have receptors for endorphins. Endorphins can antagonize FSH effects on Sertoli cells.[136,137]

Leydig-Peritubular Cell Interactions

Growth factors made by both Leydig and peritubular cells include IGF-I, TGF-α and TGF-β. Peritubular cells have androgen receptors.[138] In response to testosterone, peritubular cells secrete PModS in immature testes.[139] Thus a possible extended regulatory cascade involves luteinizing hormone stimulation of Leydig cells to make androgens which stimulate peritubular cells to make PmodS which stimulates Sertoli cells to secrete androgen binding protein.[88]

Another potential interaction of Leydig cells on peritubular cells is through Leydig production of PDGF,[140] although PDGF may be produced in other tissues and PDGF receptors are also found on Leydig cells.[141] PDGF effects on peritubular myoid cells include increase of cytosolic Ca^{++}, stimulation of proliferation and secretion of several extracellular matrix components.[140] PDGF is permissive in culture for the Leydig cell steroidogenic response to LH.[142]

Other Interactions

Other testicular somatic cell types have been investigated less than the Sertoli, Leydig and peritubular cells, but may be of importance in regulation of testicular function. For example, macrophages and Leydig cells are often found in intimate association in the interstitium, which they share with connective tissue, blood vessels and fibroblasts. Purified testicular macrophages co-cultured with Leydig cells inhibited testosterone production if LH was present in sufficient amounts.[143] The effect was mimicked by macrophage conditioned medium but not by interleukin-1 which is made by macrophages. Other candidates for macrophage cytokines acting in this system included tumor necrosis factor-α and TGF-β.

Direct Interactions

Another mode of interaction of the somatic testicular cells is by direct coupling of cells through gap junctions. There is morphological evidence for Sertoli-Sertoli, Sertoli-germ cell, and Leydig-Leydig gap junctions (see ref 144). Connexins are a family of related transmembrane proteins which form gap junctions. Connexin 43 was localized by immunoelectron microscopy to gap junctions on Sertoli and Leydig cells and dye-coupling has been demonstrated in situ between Leydig cells, and between peritubular and Sertoli cells.[144] The distribution of gap junctions depends on cell type and stage of maturation.[144] Coculture of Sertoli and peritubular cells increases synthesis of cell-associated glycosaminoglycans (carbohydrates of the proteoglycans of the ECM).[145]

SOMATIC CELL-GERM CELL INTERACTIONS

Effects of Somatic Cells on Germ Cells

A mouse gene, juvenile spermatogonial depletion (*jsd*) on chromosome 1, causes sterility and small testes as a result of blockage at the spermatogonial stage due to failure in the intratubular environment.[146,147] This environment is controlled directly or indirectly by Sertoli cells by secretion of products and interposition of the blood testis barrier. Products of Leydig or peritubular cells can interact either directly or indirectly with germ cells. A number of reviews on somatic cell-germ cell interactions are available.[103,106]

Diffusible Factors

Leydig cells secrete several molecules affecting the germ line. Testosterone is required for the completion of meiosis and for differentiation of spermatids. Testosterone by itself is not able to initiate spermatogenesis in rodents but acts synergistically with FSH. It may however be able to initiate spermatogenesis alone in primates (see ref. 148 for review). There is evidence for synergy between FSH and testosterone at the level of their receptors. FSH can stimulate androgen receptor gene expression and the amount of Sertoli receptors.[41,149] Effects may be mediated at the translational level.[41] Testosterone can stimulate FSH receptors.[150] Normal spermatogenic levels of cells can be achieved with much lower levels of testosterone than are found in the testis[151] suggesting that the testis may serve as a reservoir for supplying the rest of the body.

Leydig products other than testosterone may also be needed to maintain spermatogenesis because chemical killing of Leydig cells and supplementing with testosterone result in a decline of late spermatids.[38]

Coculture of Sertoli cells with germ cells increases the rates of DNA and RNA synthesis in the germ cells.[152] Coculture maintains glutathione synthesis[153] and elevates levels of some germ cell surface antigens.[153,154] Transferrin receptors on germ cells bind iron delivered by Sertoli cells to the adluminal compartment.[155] IL-1 which is made by Sertoli cells stimulates DNA synthesis in spermatogonia and meiotic spermatocytes.[156-158] Overexpression of IL-2 in transgenic mice causes spermatogenic arrest.[159]

Germ cells may respond to a number of factors made by Sertoli and other somatic testicular cells. In the dogfish, insulin and IGF-I stimulate

DNA synthesis of germ cells in spermatocysts (Sertoli/germ cell clones) in culture[160] suggesting that these factors stimulate Sertoli cells to produce germ cell growth factors. Germ cells have receptors for IGF-I which is made by Sertoli, Leydig and peritubular cells.[161] TGF-α receptors have not been detected on germ cells[89] and the effects of TGF-β are largely unknown.[90] Human sperm cells respond to TGF-β1 with increased synthesis of a 50 kDa oligoadenylate synthetase.[162]

Soluble recombinant Steel factor or KL normally produced by Sertoli cells stimulates DNA synthesis in Type A spermatogonia.[46] KL is stimulated in a Sertoli-mast cell coculture system by cAMP but not by testosterone, FSH, bFGF, IGF-I, TGF-α, TGF-β1, NGF, or IL-1α.[163]

Activin, which is made by Sertoli and Leydig cells, can stimulate spermatogonial proliferation.[164] Activin receptor mRNAs are expressed in spermatocytes and round spermatids. Germ cells bind activin A and inhibin A[165] and express inhibin-α and -β mRNAs stage- specifically.[166,167] In cultures of rat seminiferous tubule segments, activin A stimulated preleptotene spermatocyte and intermediate spermatogonial DNA syntheses.[168] Inhibin A inhibited DNA synthesis in both cell types. TGF-β1 had a slight stimulatory effect.

A special mode of interaction by diffusible molecules involves cycling of secretory products marked with mannose 6-phosphate (M6P).[169] Secretory products of Sertoli cells with mannose 6-phosphate recognition markers include TGF-β and procathepsin L (cysteine protease). Mannose 6-phosphate receptors (MPRs) are found in pachytene spermatocytes, round spermatids and Sertoli cells.

Such marked proteins can be delivered to lysosomes via specific receptors in the Golgi apparatus. In addition, receptors are found at the plasma membrane. Known receptors for M6P marked proteins are of two types. CI is the larger cation-independent form of about 275 kDa which is identical to the IGF-II receptor. CD is smaller, about 46 kDa, binds tightly in the presence of divalent ions, and does not bind IGF-II. Each has four domains: a signal sequence, an extracytoplasmic domain which binds ligand, a membrane spanning region and a short cytoplasmic tail lacking tyrosine kinase activity. Only CI mediates endocytosis of M6P ligands and it has been implicated in signal transduction mechanisms involving IGF-II and growth factors.

In Sertoli cells, CI-MPRs are synthesized exclusively. MPRs resembling liver CD-MPR are synthesized by germ cells.[169] In germ cells, multiple transcripts are found which may be produced by variable polyadenylation.[169] Germ cell MPRs mediate endocytosis of Sertoli products which are then proteolytically processed.[169] Pachytene spermatocytes and round spermatids cultured with a group of at least ten M6P-containing Sertoli cell secretory glycoproteins accumulated them by receptor-mediated endocytosis and processed them to lower molecular weights.[170] It has also been proposed that MPRs target hydrolytic enzymes to the acrosome of developing sperm cells.[169]

Direct Interactions

Interactions between germ cells and somatic cells may involve not only diffusible extracellular factors but may also include direct interactions mediated at the cell surfaces, diffusion of substances through gap junctions, or matrix-cell interactions. The extracellular matrix proteins fibronectin, collagen IV, and laminin are found in the basement membrane of human testis and are probably made by Sertoli cells or spermatogonia.[171] Male germ line cells except spermatogonia also express fibronectin.[171] Rat laminin-binding proteins found on Sertoli cells may mediate Sertoli-Sertoli and Sertoli-spermatogonial interactions.[172]

Integrins are an important class of adhesion molecules which exist as dimers in the plasma membrane and can interact with cytoplasmic molecules and extracellular matrix proteins. β1-Integrins were found on basement membranes[171] and associated with spermatocytes, spermatids and spermatozoa.[171,173] Integrin α-3, -5 and -6 chains are localized to the male germ line but not α-1, -2 and -4. Thus the possibilities for cell-cell or cell-matrix interactions between germ cells and basement membrane are extensive.

Liver regulating protein-like protein *(LRP)* is an integral membrane protein that regulates liver cell differentiation through direct contact and is found on Sertoli and primary spermatocyte membranes.[106,174] Antibodies to LRP block spermatocyte-stimulated secretion of transferrin by Sertoli cells. Direct transfer of material from germ cells to Sertoli cells (and vice versa) could occur through gap junctions, spermatogonial processes or tubulobulbar complexes and transfers employing pinocytotic vesicles are also possible.[175]

A glycoprotein on the surface of rat pachytene spermatocytes of 80 kDa appears to mediate binding to Sertoli cells.[176] Surfaces of rat germ cells have asialoglycoprotein (ASGP) receptors. The receptors are also found on cultured Sertoli cells and epididymal sperm.[177] The receptors have affinity for galactose. In spermatids and sperm they localize to the plasma membrane on the dorsal side of the sperm head and may be involved in spermatogenesis and sperm-egg recognitions and internalization.[177] They may bind or internalize extratesticular glycoproteins as ASGP receptors do in liver, or they could be involved in recognition of surface glycoproteins with terminal galactose moieties.[177]

Beta 1,4-Galactosyltransferase localizes to the regions of contact between primary spermatocytes and Sertoli cells by indirect immunofluorescence.[178] Antibody treatment of primary spermatocytes but not Sertoli cells inhibits cell adhesion. *Fucosyltransferase* found on Sertoli cell surfaces may also function in cell adhesion to germ cells.[179] An approximately 30 kDa Sertoli cell membrane protein may bind spermatocytes, spermatids and residual bodies to Sertoli cells.[180]

Gap junctions have been detected between Sertoli cells and germ cells. There is evidence for electrical coupling between cells of the seminiferous epithelium and metabolic coupling between pachytene spermatocytes and Sertoli cells.[144] Desmosome-like junctions and ectoplasmic specializations also connect Sertoli cells and germ cells.[181] For a review of junctions between germ cells and Sertoli cells, see ref 175.

Effects of Germ Cells on Somatic Cells

Several approaches have been taken to demonstrate effects of germ cells on testicular somatic cell functions. Experimental manipulation of the types of germ cells present by chemical treatments, irradiation, heat, or mutation allow correlations of somatic cell function with germ cell subtypes. However, each treatment is likely to have multiple secondary effects on tissues that make interpretation sometimes difficult. Coculture of somatic cells in the presence or absence of germ cells has also been exploited to elucidate the effects of germ cells. Unfortunately, culture systems are still somewhat primitive and conditions are likely to differ from those in whole animals. Alternatively, medium conditioned by germ cells has been shown to have several effects on Sertoli cells. Again, one must be aware that in vitro systems do not always behave as the corresponding in vivo systems.

Clearly some somatic functions are not dependent upon germ cells. For example, in the absence of germ cells, Sertoli cells can differentiate and give rise to a germ-cell free seminiferous epithelium.[182] Sertoli cells can produce anti-Mullerian hormone in the absence of germ cells in W^e/W^e mice.[183]

Germ Cell Depletion

Depletion of selected classes of germ cells demonstrates that Sertoli cells are controlled positively or negatively by the presence of particular germ cells. Germ cells can be depleted by a variety of agents such as γ-rays, heat, X-rays, and chemicals such as bisulfan, procarbazine and metoxyacetic acid. In this regard, the age of the animal may be important, since the modes of regulation may vary with development. It has been proposed that the most advanced germ cells present in the seminiferous epithelium at various times of development may regulate Sertoli function.[184] For example, in mature rats, late spermatids are required for normal Sertoli function.[175,184] In immature rats, loss of early spermatids and pachytene spermatocytes lowers ABP and FSH production,[185] but in adult rats their levels are constant.[186,187]

Germ cells affect overall protein secretion by isolated rat seminiferous epithelia in a complex way dependent upon stage.[188] The protein secretory increase seen at Stages VI-VIII is dependent on testosterone, pachytene spermatocytes and round spermatids. Depletion of only one germ cell type had no effect at Stages I-V and only depletion of round spermatids affected Stages IX-XIV. Not all secreted proteins were similarly affected; SGP-1 and SGP-2 were relatively insensitive to germ cells.

Late spermatids appear to be required for regulation of many Sertoli proteins. Decreases in ABP,[186] inhibin,[99,186] transferrin,[189] FSH receptors[190] and increases in testins[191] have been reported for late spermatid-depleted testes.

Cyclic protein-2 transcripts increase when pachytene spermatocytes or round spermatids are chemically depleted and decrease when elongated spermatids are depleted, suggesting that elongated spermatids positively modulate, and earlier germ cells negatively regulate CP-2 in adult rats.[192] In prepubertal rats, round spermatids have been reported to stimulate CP-2 production.[193]

In irradiated testis, preleptotene spermatocytes have been shown to regulate plasminogen activator secretion.[194] Pachytene spermatocytes also seem to regulate inhibin production and gene expression,[186,195] inhibit

FSH[190] and stimulate ABP.[195] Depletion of germ cells, especially spermatocytes and early spermatids, lowers production of many secreted proteins from seminiferous tubules at stages VI-VIII.[196] Depletion of germ cells in vivo by X-irradiation also causes increases of testicular testin.[197]

In adult rats, heat lowers pachytene spermatocyte and early spermatid numbers. When late spermatids were decreased as a consequence, a decline of ABP was seen which was reversed when full spermatogenesis resumed.[185] A similar phenomenon occurs during post-irradiation recovery when the late spermatids appear and Sertoli activities resume.[186]

Coculture Systems

Although differentiation of the male germ line cells in culture is limited, coculture of Sertoli cells with enriched fractions of germ cells can be used to assess effects of germ cells on Sertoli cells. In general, where data are available, the effects monitored in coculture reflect those in vivo.[175] Again the effects may depend on age of the Sertoli cells.[198-299] Early spermatids regulate estradiol and ABP of older Sertoli cells compared to pachytene spermatocytes whose effects are less age-dependent.[198,199]

In coculture, germ cells, especially pachytene spermatocytes, stimulate production of transferrin[199-201] and transferrin mRNA,[202] ABP,[200,203] inhibin and inhibin mRNA,[204] γ-GTP,[205] preproenkephalin[206] and protein phosphorylation[207] and inhibit production of estradiol[198,199,208] and Leydig steroidogenesis stimulator[209] by Sertoli cells. The effect of germ cells on protein phosphorylation in Sertoli cells involves intracellular Ca^{++} signaling via the inositol triphosphate pathway.[207]

Spermatids also stimulate production of ABP,[210] transferrin[199] and inhibin[204] and levels of proenkephalin mRNA[206] and inhibit estradiol production[198,199] in coculture. Residual bodies stimulate production of IL-1α and IL-6 by Sertoli cells.[211,212] Since residual bodies are phagocytosed by Sertoli cells, they have been suggested to influence Sertoli cells directly by mRNA ingestion or by the production of cytokines in response to the phagocytic process.[184]

The mRNA encoding the inhibin-α subunit increases when Sertoli cells are cocultured with round spermatids from 20 day rats.[204] Secretion of α-inhibin decreased in coculture with elongated but not round spermatids from adult rats.[195] Addition of isolated pachytene spermatocytes or early spermatids to Sertoli cells stimulated inhibin production.[208]

Of 21 secreted Sertoli cell products in coculture with germ cells, one was stimulated and two were inhibited by all cell types[213]. Early spermatids and residual bodies enhanced one and inhibited two as well as inducing a new secretory protein of 72 kDa called GIP.

An indirect effect of germ cells on Leydig function via Sertoli cells has been investigated in a bicameral culture system in which cells are separated by a porous membrane. Sertoli cells secrete Leydig steroidogenesis proteins which can be inhibited by added pachytene spermatocytes.[209]

Conditioned Medium

Another approach is to test conditioned medium obtained from germ cell cultures on Sertoli cells. Conditioned medium from germ cells[214] has been shown to lower Sertoli cell RNA synthesis[152] and stimulate phosphorylation of various proteins including transferrin.[199,202,215-217] It enhances production of γ-GTP,[205,218] SGP-1 and ceruloplasmin,[217,202] and accumulation of mRNAs for inhibin,[204] transferrin,[202] and pre-proenkaphalin.[206] Germ cell conditioned medium inhibits 17β-estradiol production[198,199,210] and Leydig cell steroidogenesis factor.[209] Secretion of the α-subunit of inhibin is increased in GC conditioned medium.[204,205]

Germ cell conditioned medium increases ABP synthesis and secretion.[210,219,220] Isolated pachytene spermatocytes or early spermatids stimulate isolated Sertoli cells to produce ABP,[208] and late spermatids are essential for ABP production.[99,186,187,221] Germ cell conditioned medium stimulates transferrin production and secretion.[205,216,217,220,222]

Conditioned medium from germ cells has been reported to inhibit or stimulate SGP-2[199,221] or testin production.[217,222,221] The discrepancy may be explained by the degree of germ cell viability in culture.[175] SGP-1 and SGP-2 mRNA levels are independent of germ cells in coculture or in vivo, and changes seen in conditioned medium are likely to reflect post-transcriptional events.[202,233]

Testin secretion is inhibited by germ cell conditioned medium.[197,221] Testin secretion appears to be negatively regulated especially by late spermatids. This raises the possibility that late spermatids may alter their junctional attachments to Sertoli cells by decreasing testin secretion facilitating their release. On the other hand, other workers have reported stimulation of testin secretion by germ cells.[217]

Purification of Factors

Several attempts have been made to purify some of the active components of germ cell conditioned medium that affect somatic cells.[198,199,202,206,214,216,217,221,222] Conditioned medium from round spermatids was concentrated, purified and tested for stimulation of total protein secretion and transferrin secretion from isolated Sertoli cells.[214] Two proteins of 29 kDa and 24.5 kDa, not isolable from other germ cell conditioned media or somatic testicular cells, were obtained. The N-terminal amino acid sequences were not homologous with any known proteins.

Several factors which modulate specific Sertoli secretions have been purified from pachytene spermatocyte or early spermatid cultures.[221] Two protein fractions inhibiting testin secretion, three inhibiting clusterin secretion, and four stimulating transferrin secretion using ion-exchange and gel permeation chromatography and reversed phase HPLC were reported. These are as yet incompletely characterized.

A protein of > 30,000 kDa stimulates preproenkephalin gene expression in Sertoli cells.[206] Conditioned medium from round spermatids containing this factor stimulates preproenkephalin mRNA levels by 7.6-fold within 12 hours.

Some specific products that modify Sertoli or Leydig cell functions are acrogranin, NGF, bFGF, and tumor necrosis factor. Acrogranin is an acrosomal marker, a glycoprotein of 67 kDa found from pachytene spermatocytes to mature sperm.[224] Immunolocalization suggests that the protein is testis-specific but its mRNA displays a ubiquitous expression.

Primary structures predicted from the cDNA sequence show it to be a member of a family of cysteine-rich molecules homologous to human rat epithelin/granulin peptides. These peptides modulate growth and acrogranin appears to be a precursor to these potential growth factors in the testis. It has been proposed to signal Sertoli cells.[224]

NGF and bFGF are found in germ cell conditioned medium. NGF protein and mRNA are expressed in spermatocytes and early spermatids.[225] NGF receptors are present in Sertoli cells and their expression is regulated by androgens.[132] That Sertoli cells respond to the growth factor in vivo is shown by infusion of β-NGF into rat testis which increases ABP mRNA in Sertoli cells and prolongs Stages VII-VIII. β-NGF acts by increasing the number of Sertoli cells active in ABP mRNA synthesis.[226]

Germ cells, primarily pachytene spermatocytes, appear to be the main source of bFGF in the testis.[227,228] Sertoli cells in vitro can respond to bFGF by up regulation of receptors of FSH, by making plasminogen activator[229] or by synthesis of *c-fos* mRNA.[230-232] bFGF may stimulate proliferation of prepubertal Sertoli cells.[229,233] bFGF purified from germ cell conditioned medium[124] stimulates transferrin mRNA levels in Sertoli cells. bFGF has been reported to stimulate testosterone synthesis in Leydig cells.[235] hCG, IGF-I, bFGF and EGF all stimulate *c-fos* expression in cultured Leydig cells.[125]

Tumor necrosis factor TNF-α is a 235-amino acid 17.3 kDa protein made as a 26 kDa precursor which is secreted by many cell types. It has broad effects on cell growth and differentiation. TNF-α mRNA is found in pachytene spermatocytes and round spermatids of 1.9 and 2.8 kb respectively.[236] Highest levels of its mRNA are found in Stages IX-XI.[236] Small amounts of TNF are detected in round spermatid conditioned medium. TNF has been proposed as a paracrine factor[236] but this is yet uncertain. Infusion of TNF into rats affects Leydig cell secretion of testosterone and results in loss of spermatocytes and spermatids.[237] TNF affects Sertoli cells[238] by stimulating transferrin release Almost all cells have TNF receptors. Two receptors A and B with different cytoplasmic domains suggest different pathways of signal transduction. In the testis, receptor mRNA is localized to Sertoli cells but not pachytene spermatocytes or round spermatids.

Other factors are not as well characterized. The mRNA for a growth hormone releasing hormone-like factor from testis *(t-GHRH-li)* is expressed highly in round spermatids and spermatocytes, little in Sertoli cells, and not at all in elongating spermatids, Leydig cells or peritubular cells.[239] t-GHRH-li is synthesized in early germ cells, has no effect on Leydig steroidogenesis but is capable of stimulating GH secretion from the anterior pituitary in rats. No testicular receptor has yet been identified, but it has been suggested that t-GHRH-li may act on Sertoli cells. Application of GHRH to Sertoli cells stimulates cAMP levels and *c-fos* and steel factor mRNA production with no effect on SGP-2 mRNA[240] suggesting a Sertoli testicular target as well as a pituitary target for this hormone. EGF from post-meiotic germ cells[241] and inhibin from spermatogonia[242] are also candidates for regulation of Sertoli cells.

SUMMARY

Control of the spermatogenic program involves several regulatory circuits and interactions of various tissues and cell types. The testis-determining factor is currently attributed to the *SRY* gene. The *c-kit*/KL interaction provides a basis for control of spermatogonial stem cell proliferation. The pituitary and testis govern each others' activities. Somatic cells of the testis interact in autocrine and paracrine loops. Testicular somatic cells control the progress of spermatogenesis, but germ cells also affect the activities of the somatic cells and ultimately of their own progression.

REFERENCES

1. Schroeder, P.C. Endogenous control of gametogenesis. In: Giese, A.C., Pearse, J.S. and Pearse, V.B., eds. Reproduction of Marine Invertebrates. Palo Alto: Blackwell/Boxwood, 1987:179-249.
2. Palmer, S.J. and Burgoyne, P.S. In situ analysis of fetal, prepubuertal and adult XX/XY chimaeric mouse testes: Sertoli cells are predominantly, but not exclusivley, XY. Development 1989; 112:265-268.
3. Harley, V.R. Genetic control of testis determination. In: deKretser, D. Molecular Biology of the Male Reproductive System. NY: Acad. Press, 1993:3-20.
4. Capel, B. and Lovell-Badge, R. The Sry gene and sex determination in mammals. In: Wassserman, P. ed. Advances in Developmental Biology, Vol 2. Greenwich, CT: A.I. Press Inc., in press.
5. Sinclair, A.H., Berta, P., Palmer, M.S., Hawkins, J.R., Griffiths, B.L., Smith, M.J., Foster, J.W., Frischauf, A.-M., Lovell-Badge, R. and Goodfellow, P.N. A gene from the human sex-determining region encodes a protein with homology to a conserved DNA-binding motif. Nature 1990; 346:240-244.
6. Gubbay, J., Collignon, J., Koopman, P., Capel, B., Economou, A., Munsterberg, A., Vivian, N., Goodfellow, P.N. and Lovell-Badge, R. A gene mapping to the sex-determining region of the mouse Y chromosome is a member of a novel family of embryonically expressed genes. Nature 1990; 346:245-250.
7. Nasrin, N., Buggs, C., Kong, X.F., Carnazza, J., Goebl, M. and Alexander-Bridges, M. DNA binding properties of the product of the testis-determining gene and a related protein. Nature 1991; 354:317-320.
8. Harley, V.R., Jackson, D.I., Hextall, P., Hawkins, J.R., Berkovitz, G.D., Sockanathan, S., Lovell-Badge, R. and Goodfellow, P.N. DNA binding activity of a recombinant SRY from normal males and XY females. Science 1992; 255:453-456.
9. Whitfield, L.S., Lovell-Badge, R. and Goodfellow, P.N. Rapid sequence evolution of the mammalian sex-determining gene SRY. Nature 1993; 364:713-715.
10. Tucker, P.K. and Lundrigan, B.L. Rapid evolution of the sex determining locus in Old World mice and rats. Nature 1993; 364:715-717.
11. Denny, P., Swift, S., Connor, F. and Ashworth, A. An SRY-related gene expressed during spermatogenesis in the mouse encodes a sequence-specific DNA-binding protein. EMBO J. 1992; 11:3705-3712.

12. Thompson, J.N., Howell, J., McC. and Pitt, G.A.J. Vitamin A and reproduction in rats. Proc. Roy. Soc. B Biol. Sci. 1964; 159:510-535.
13. Griswold, M.D., Bishop, P.D., Kim, K.H., Ren, P., Sliteri, K.E. and Morales, C. Function of vitamin A in normal and synchronized seminiferous tubules. Ann. N.Y. Acad. Sci. 1989; 564:154-172.
14. Wang, Z. and Kim, K.H. Vitamin A-deficient testis germ cells are arrested at the end of S phase of the cell cycle: A molecular study of the origin of synchronous spermatogenesis in regenerated seminiferous tubules. Biol. Reprod. 1993; 48:1158-1165.
15. Wagner, E.F. and Alexander, W.W. Of *kit* and mouse and man. Curr. Biol 1991; 1:356-358.
16. Besmer, P. The *kit* ligand encoded at the murine *Steel* locus: a pleiotropic growth and differentiation factor. Curr. Opin. Cell Biol. 1991; 3:939-946.
17. Witte, O.N. *Steel* locus defines new multipotent growth factor. Cell 1990; 63:5-6.
18. Keshet, E., Lyman, S.D., Williams, D.E. anderson, D.M., Jenkins, N.A. Copeland, N.D. and Parada, L.F. Embryonic expression patterns of the *c-kit* receptor and its cognate ligand suggest multiple functional roles in mouse development. EMBO J. 1991; 10:2425-2435.
19. Yoshinaga, K., Nishikawa, S., Ogawa, M., Hayashi, S.I., Kunisada, T., Fujimoto, T. and Nishikawa, S.I. The role of *c-kit* in mouse spermatogenesis: Identification of spermatogonia as a specific site of *c-kit* expression and function. Development 1991; 113:689-699.
20. Orr-Urtreger, A., Avivi, A., Zimmer, Y., Givol, D., Yarden, Y. and Lonai, P. Developmental expression of *c-kit,* a proto-oncogene encoded by the *W* locus. Development 1990; 109:911-923.
21. Manova, K., Nocka, K., Besmer, P. and Bachvarova, R.F. Gonadal expression of *c-kit* encoded at the *W* locus of the mouse. Development 1990; 110:1057-1069.
22. Manova, K. and Bachvarova, R.F. Expression of *c-kit* encoded at the *W* locus of mice in developing embryonic germ cells and presumptive melanoblasts. Devel. Biol. 1991; 146:312-324.
23. Dolci, S., Williams, D.E., Ernst, M.K., Resnick, J.L., Brannan, C.I., Lock, L.F., Lyman, S.D., Boswell, H.S. and Donovan, P.J. Requirement for mast cell growth factor for primordial germ cell survival in culture. Nature 1991; 352:809-811.
24. Flanagan, J.G., Chan, D.C. and Leder, P. Transmembrane form of the *kit* ligand growth factor is determined by alternative splicing and is missing in the Sl^d mutant. Cell 1991; 64:1025-1035.
25. Marziali, G., Lazzaro, D. and Sorrentino, V. Binding of germ cells to mutant Sl^d Sertoli cells is defective and is rescued by expression of the transmembrane form of the *c-kit* ligand. Devel. Biol. 1993; 157:182-190.
26. Geremia, R., Sorrentino, V. and Rossi, P. Oncogene expression in male germ cells at various stages of differentiation. In: Baccetti, B. ed. Comparative Spermatology 20 Years After. NY: Raven, 1991:723-726.
27. Sorrentino, V., Giorgi, M., Geremia, R., Besmer, P. and Rossi, P. Expression of the *c-kit* proto-oncogene in the murine male germ cells. Oncogene 1991; 6:149-151.
28. Rossi, P., Marziali, G, Albanesi, C., Charlesworth, A., Geremia, E. and Sorrentino, V. A germ cell specific promoter creates a novel *c-kit* tran-

script. Devel. Biol. 1992; 152:203-207.

29. Byskov, A.G. and Saxen, L. Induction of meiosis in fetal mouse testis in vitro. Devel. Biol. 1976; 52:193-200.
30. Byskov, A.G., Hoyer, P.E., Bjorkstrom, N., Mork, A.B., Olsen, B. and Grinsted, J. Ultrastructure of germ cells and adjacent somatic cells correlated to initiation of meiosis in the fetal pig. Anat. Embryol. 1986; 175:57-67.
31. Taketo, T., Merchant-Larios, H. and Koide, S.S. Induction of testicular differentiation in the fetal mouse ovary by transplantation into adult male mice. Proc. Soc. Exp. Biol. Med. 1984; 176:148-153.
32. Vaishnav, M.Y. and Moudgal, R.N. Effects of specific FSH deprivation on testicular germ cell transformations and on LDH-X hyaluronidase activity of immature and adult rats. In: Hunzicker-Dunn M. and Schwartz, N.B. eds. Follicle Stimulating Hormone - Regulation of Secretion and Molecular Mechanism of Action. New York: Springer-Verlag, 1992: 364-368.
33. Orth, J.M. The role of follicle-stimulating hormone in controlling Sertoli cell proliferation in the testes of fetal rats. Endocrinol. 1984; 115: 1248-1255.
34. Dym, M., Raj, H.G.M., Lin, Y.C., Chemes, H.E., Kotite, N.J., Nayfeh, S.N. and French, F.S. Is FSH required for maintenance of spermatogenesis in adult rats? J. Reprod. Fertil. 1978; 17:175-181.
35. Woods, M.C. and Simpson, M.E. Pituitary control of the testis of the hypophysectomized rat. Endocrinol. 1961; 69:91-125.
36. Bartlett, J.M.S., Weinbauer, G.F. and Nieschlag, E. Differential effects of FSH and testosterone on the maintenance of spermatogenesis in the adult hypophysectomized rat. J. Endocrinol. 1989; 121:49-58.
37. Weinbauer, G.F., Behre, H.M., Fingscheidt, U. and Nieschlag, E. Human follicle stimulating hormone exerts a stimulating effect on spermatogenesis, testicular size and serum inhibin levels in the gonadotropin-releasing hormone antagonist-treated nonhuman primate *Macaca fascicularis.* Endocrinol. 1991; 129:1831-1839.
38. Roberts, K.P. and Zirkin, B.R. Androgen regulation of spermatogenesis in the rat. Ann. N.Y. Acad. Sci. 1991; 637:90-106.
39. Mullaney, B.P. and Skinner, M.K. Transforming growth factor-β b1, b2, and b3 gene expression and action during pubertal development of the seminiferous tubule: Potential role at the onset of spermatogenesis. Molec. Endocrinol 1993; 7:67-76.
40. Spiteri-Grech, J. and Nieschlag, E. The role of growth hormone and insulin-like growth factor I in the regulation of male reproductive function. Horm. Res. 1992; 38 Supp 1:22-27.
41. Blok, L.J., Hoogerbrugge, J.W., Themmen, A.P.N., Baarends, W.M., Post, M. and Grootegoed, J.A. Transient down-regulation of androgen receptor messenger ribonucleic acid mRNA expression in Sertoli cells by follicle-stimulating hormone is followed by up-regulation of androgen receptor mRNA and protein. Endocrinol. 1992; 131:1343-1349.
42. Shubhada, S., Glinz, M. and Lamb, D.J. Sertoli cell secreted growth factor. Cellular origin, paracrine and endocrine regulation of secretion. J. Androl. 1993; 14:99-109.
43. Lamb, D.J. Growth factors and testicular development. J. Urol. 1993; 150:583-592.

44. Lamb, D.J., Spotts, G.S, Shubhada, S. and Baker, K.R. Partial characterization of a unique mitogenic activity secreted by Sertoli cells. Mol.Cell. Endocrinol. 1991; 79:1-12.
45. Rossi, P., Dolci, S., Albanesi, C., Grimaldi, P., Ricca, R. and Geremia, R. Follicle-stimulating hormone induction of steel factor SLF mRNA in mouse Sertoli cells and simulation of DNA synthesis in spermatogonia by soluble SLF. Devel. Biol. 1993; 155:68-74.
46. Ying, S.-Y. and Murata, T. Molecular biology of the gonadal proteins-inhibin, activin, and follistatin. In: Gwatkin, R.G.B. ed. Genes in Mammalian Reproduction. New York: Wiley-Liss, 1993:207-228.
47. Namiki, M., Kondoh, N., Matsumiya, K., Sakoda, S., Nishimune, Y. and Okuyama, A. Hormonal regulation of human testicular inhibin α and βB subunit messenger RNAs. J. Urol. 1993; 150:1007-1009.
48. Kangasniemi, M. Effects of adenosine analog PIA n-phenylisopropyladenosine on FSH-stimulated cyclic AMP cAMP production in the rat seminiferous epithelium. Molec. Cell. Endocrinol. 1993; 96:141-146.
49. Foulkes, N.S., Scholtter, F., Pevet, P. and Sassone-Corsi, P. Pituitary hormone FSH directs the CREM functional switch during spermatogenesis. Nature 1993; 362:264-267.
50. Foulkes, N.S., Mellstrom, B., Benusiglio, E. and Sassone-Corsi, P. Developmental switch of CREM function during spermatogenesis: from antagonist to activator. Nature 1992; 355:80-84.
51. Schlatt, S., Weinbauer, G.F., Arslan, M. and Nieschlag, E. Appearance of α-smooth muscle actin in peritubular cells of monkey testes is induced by androgens, modulated by follice-stimulating hormone, and maintained after hormone withdrawal. J. Androl. 1993; 14:340-351.
52. Huhtaniemi, I.T. and Pelliniemi, L.J. Fetal Leydig cells: Cellular origin, morphology, life span and special functional features. Proc. Soc. Exp. Biol. Med. 1992; 201:125-140.
53. Jost, A. Hormonal factors in the sex differentiation of the mammalian foetus. Philos. Trans. R. Soc. London. B. 1970; 259:119-130.
54. Huhtaniemi, H. Hormonal control mechanisms of Leydig cells. In: deKretser, D. ed. Molecular Biology of the Male Reproductive System. NY: Acad. Press, 1993:383-410.
55. Tres, L.L., Smith, E.P., van Wyk, J.J. and Kierszenbaum, A.L. Immunoreactive sites and accumulation of somatomedin-C in rat Sertoli-spermatogenic cell co-cultures. Exp. Cell Res. 1986; 162:33-50.
56. Perrard-Sapori, M.H., Chatelain, P.C., Jaillard, C. and Saez, J.M. Charactrerization and regulation of somatomedin C/insulin-like growth factor I SmC/IGF-I receptors on cultured pig Leydig cells. Effects of SmC/IGF-I on luteotropin receptor and steroidogenesis. Eur. J. Biochem. 1987; 165:209-214.
57. Nagpal, M.L., Wang, D., Calkins, J.H., Chang, W. and Lin, T. Human chorionic gonadotropin up-regulates insulin-like growth factor-I receptor gene expression of Leydig cells. Endocrinol. 1991; 129:2820-2826.
58. Benahmed, M., Morera, A.M., Chauvin, M.C. and de Peretti, E. Somatomedin-c/insulin-like growth factor 1 as a possible intratesticular regulator of Leydig cell activity. Mol. Cell. Endocrinol. 1987; 50:69-74.
59. Cooke, B.A. Platts, E.A., Abayasekera, D.R.E. and Rose, M.P. Mechanisms of hormone-induced densensitization of adenylate cyclase. Biochem.

Soc. Trans. 1991; 17:633-635.
60. Catt, K.J., Harwood, J.P., Clayton, R.N., Davies, T.F., Chan, V., Katikinemi, M., Nozu, K. and Dufau. M.L. Regulation of peptide hormone receptors and gonadal steroidogenesis. Rec. Prog. Horm. Res. 1980; 36:557-622.
61. Meidan, R. and Hseuh, A.J.W. Identification and characterization of vasopressin receptors in the rat testis. Endocrinol. 1985; 116:416-423.
62. Rigaudiere, N., Loubassou, S., Grizard, G. and Boucher, D. Characterization of insulin binding and comparative action of insulin and insulin-like growth factor I on purified Leydig cells from the adult rat. Int. J. Androl. 1988; 11:165-178.
63. Mukhopadhyay, A.K., Helbing, J. and Leidenberger, F.A. The role of Ca^{++} and calmodulin in the regulation of atrial natriuretic factor in isolated mouse Leydig cells is mediated by cyclic GMP. Endocrinol. 1989; 125:686-692.
64. Lobie, P.E., Breipohl, W., Garcia-Aragon, J. and Waters, M.J. Cellular localization of the growth hormone receptor/binding protein in the male and female reproductive systems. Endocrinol. 1990; 126:2214-2221.
65. Tapanainen, J.S., Tilly, J.L., Vihko, K.K. and Hsueh, A.J.W. Hormonal control of apototic cell death in the testis: Gonadotropins and androgens as testicular cell survival factors. Molec. Endocrinol. 1993; 7:643-650.
66. Sinha-Hikim, A.P. and Swerdloff, R.S. Temporal and stage-specific changes in spermatogenesis of rat after gonadotropin deprivation by a potent gonadotropin-releasing hormone antagonist treatment. Endocrinol. 1993; 133:2161-2170.
67. Weinbauer, G.F. and Nieschlag, E. Gonadotropin-releasing hormone analogue-induced manipulation of testicular function in the monkey. Human Reprod. 1993; 8 Suppl. 2:45-50.
68. Muffly, K.E., Nazian, S.J. and Cameron, D.F. Junction-related Sertoli cell cytoskeleton in testosterone-treated hypophysectomized rats. Biol. Reprod. 1993; 49:1122-1132.
69. Pogach, L., Giglio, W., Nathan, E. and Huang, H.F.S. Maintenance of spermiogenesis by exogenous testosterone in rats treated with a GnRH antagonist: relationship with androgen-binding protein status. J. Reprod. Fert. 1993; 98:415-422.
70. Ling, N., Ying, S.-Y., Ueno, N., Esch, F., Denoroy, L. and Guillemin, R. Isolation and partial characterization of a Mr 32,000 protein with inhibin activity from porcine follicular fluid. Proc. Natl. Acad. Sci. USA 1985; 82:7217-7221.
71. Robertson, D.M., Foulds, L., Leversha, L., Morgan, F.J., Hearn, M.T.V., Burger, H.G., Wettenhall, R.E.H. and deKretser, D.M. Isolation of inhibin from bovine follicular fluid. Biochem. Biophys. Res. Comm. 1985; 126:220-226.
72. Tindall, D.J., Rowley, D.R., Murthy, L., Lipshultz, L.I. and Chang, C.H. Structure and biochemistry of the Sertoli cell. Int. Rev. Cytol. 1985; 94:127-149.
73. De Jong, F.H. Inhibin. Physiol. Rev. 1988; 68:555-607.
74. Vaughn, J.M., Rivier, J., Corrigan, A.Z., McClintock, R., Campen,, C.A., Jolley, D., Voglmayr, J.K., Bardin, C.W., Rivier, C. and Vale, W. Inhibin. Meth. Enzymol. 1989; 168:588-617.

75. deWinter, J.P., Themmen, A.P.N., Hoogerbrugge, J.W., Klaij, L.A., Grootegoed, J.A. and de Jong, F.H. Activin receptor mRNA expression in rat testicular cell types. Molec. Cell. Endocrinol. 1992; 83:R1-R8.
76. Grootenhuis, A.J., Steenbergen, J., Timmerman, M.A., Dorsman, A.N.R.D., Schaaper, W.M.M., Meloen, R.H. and deJong, F.H. Activin. J. Endocrinol. 1989; 122:293-301.
77. Kaipia, A., Penttila, T.-L., Shimasaki, S., Ling, N. Parvinen, M. and Toppari, J. Expression of inhibin β-A and β-B, follistatin and activin-A receptor messenger ribonucleic acids in the rat seminiferous epithelium. Endocrinol. 1992; 131:2703-2710.
78. Attardi, B., Keeping, H.S., Winter, S.J., Kotsuji, F., Maurer, R.A. and Troen, P. Rapid and profound suppression of messenger ribonucleic acid encoding follicle-stimulating hormone β by inhibin from primate Sertoli cells. Mol. Endocrinol. 1989; 3:280-287.
79. Sealfon, S.C., Laws, S.C., Wu, J.C., Gillo, B. and Miller, W.L. Hormonal regulation of gonadotropin-releasing hormone receptors and messenger RNA activity in ovine pituitary culture. Endocrinol. 1990; 4:1980-1987.
80. Attardi, B., Vaughn, J. and Vale, W. Regulation of FSHβ messenger ribonucleic acid levels in the rat by endogenous inhibin. Endocrinol. 1991; 129:2802-2804.
81. Billestrup, N., Gonzalez-Manchon, C., Potter, E. and Vale, W. Inhibition of somatotroph growth and growth hormone biosynthesis by activin in vitro. Mol. Endocrinol. 1990; 4:356-362.
82. Vale, W., Rivier, C., Hseuh, A., Campen, C., Meurnier, H., Bicsak, T., Vaughn, J., Corrigan, A., Bardin, W., Sawchecko, P., Petraglia, F., Plotsky, P., Spiess, J. and Rivier, C. Chemical and biological characterization of the inhibin family of protein hormones. Recent Prog. Horm. Res. 1988; 44:1-34.
83. Kitaoka, M., Kojima, I. and Ogata, E. Activin A: A modulator of multiple types of anterior pituitary cells. Biochem. Biophys. Res. Comm. 1988; 157:48-54.
84. Vale, W., Rivier, J., Vaughan, J., McClintock, T., Corrigan, A., Woo, W., Karr, D. and Spiess, J. Purification and characterization of an FSH releasing protein from porcine ovarian fluid. Nature 1986; 321:776-779.
85. Murata, T. and Ying, S.Y. Transforming growth factor-β and activin inhibit basal secretion of prolactin in a pituitary extract. Proc. Soc. Exp. Biol. Med. 1991; 198:599-605.
86. Roberts, K.P., Santulli, R., Seiden, J. and Zirkin, B.R. The effect of testosterone withdrawal and subsequent germ cell depletion on transferrin and sulfated glycoprotein-2 messenger ribonucleic acid levels in the adult testis. Biol. Reprod. 1992; 47:92-96.
87. Fanjul, L.F., Gonzalez, J., Quintana, J., Santana, P., Hernandez, I., Cabrere, J., Estevez, F. and Ruiz de Galarreta, C.M. Inhibition of steroidogenesis in neonatal Leydig cells by unknown factors present in spent media of androgen-treated cultured testicular cells. J. Androl. 1993; 14:419-427.
88. Skinner, M.K., Norton, J.N., Mullaney, B.P., Rosselli, M., Whaley, P.D. and Anthony, C.T. Cell-cell interactions and the regulation of testis function. Ann. N.Y. Acad. Sci. 1991; 637:354-363.

89. Skinner, M.K., Takacs, K. and Coffey, R.J. Cellular localization of transforming growth factor-α gene expression and action in the seminiferous tubule: Peritubular cell-Sertoli cell interactions. Endocrinol. 1989; 124:845-854.
90. Skinner, M.K. and Moses, H.L. Transforming growth factor-β gene expression and action in the seminiferous tubule: peritubular cell-Sertoli cell interactions. Mol. Endocrinol. 1989; 3:625-634.
91. Hutson, J.C. and Stocco, C.M. Peritubular cell influence on the efficiency of androgen-binding protein secretion by Sertoli cells in culture. Endocrinol. 1981; 108:1362-1368.
92. Suarez-Quian, C.A., Dai, M., Onoda, M., Kriss, R.M. and Dym, M. Epidermal growth factor receptor localization in the rat and monkey testis. Biol. Reprod. 1989; 41:921-932.
93. Nargowalla, C., McCabe, D. and Fritz, I.B. Modulation of levels of messenger RNA for tissue-type plasminogen activator in rat Sertoli cells, and levels of messenger RNA for plasminogen activator inhibitor in peritubular cells. Molec. Cell. Endocrinol. 1990; 70:73-80.
94. Hettle, A.J., Balekjian, E., Tung, P.S. and Fritz, I.B. Rat testicular peritubular cells in culture secrete an inhibitor of plasminogen activator activity. Biol. Reprod. 1988; 38:359-371.
95. Ailenberg, M., Tung, P.S. and Fritz, I.B. Transforming growth factor-β elicits shape changes and increases contractility of testicular peritubular cells. Biol. Reprod. 1990; 42:499-509.
96. Saez, J.M., Chatelain, P., Mather, J.P. and Saez, J.M. Differentiating effects of somatomedin-C/insulin-like growth factor I and insulin on Leydig and Sertoli cell functions. Reprod. Nutr. Devel. 1988; 28:989-1008.
97. Spaliviero, J.A. and Handelsman, D.A. Effect of epidermal growth factor and insulin-like growth factors on vectorial secretion of transferrin by rat Sertoli cells in vitro. Mol. Cell. Endocrinol. 1991; 81:95-104.
98. Smith, E.P., Dickson, B.A. and Chernausek, S.D. Insulin-like growth factor binding protein-3 secretion from cultured rat Sertoli cells: dual regulation by follicle stimulating hormone and insulin-like growth factor-I. Endocrinol. 1990; 127:2744-2755.
99. Pinon-Latillade, G., Vignier-Martinez, M.C. and Maas, J. Influence of germ cells upon Sertoli cells during continous low-dose rate γ-irradiation of adult rats. Mol. Cell. Endocrinol. 1985; 58:51-63.
100. Verhoeven, G. and Cailleau, J.A. Testicular peritubular cells secrete a protein under androgen control that inhibits induction of aromatase activity in Sertoli cells. Endocrinol. 1988; 123:2100-2110.
101. Norton, J.N. and Skinner, M.K. Regulation of Sertoli cell differentiation by the testicular paracrine factor PModS: Potential role of immediate-early genes. Molec. Endocrinol. 1992; 6:2018-2026.
102. Skinner, M.K. Cell-cell interactions in the testis. Endocrinol. Rev. 1991; 12:45-77.
103. Sharpe, R.M. Experimental evidence for Sertoli-germ cell and Sertoli-Leydig cell interactions. In: Russell L.D. and Griswold, M.D. eds. The Sertoli Cell. Clearwater, FL: Cache River Press, 1993:391-419.
104. Bellve, A.R. and Zheng, W. Growth factors as autocrine and paracrine modulators of male gonadal functions. J. Reprod. Fertil. 1989; 85:771-793.
105. Ojeifo, J.O., Byers, S.W., Papadopoulos, V. and Dym, M. Sertoli cell-

secreted proteins stimulates DNA synthesis in purified rat Leydig cells in vitro. J. Reprod. Fertil. 1990; 90:93-108.

106. Jegou, B. and Sharpe, R.M. Paracrine mechanisms in testicular control. In: deKretser, D. ed. Molecular Biology of the Male Reproductive System. NY: Acad. Press, 1993:271-310.

107. Rosselli, M. and Skinner, M.K. Developmental regulation of Sertoli cell aromatase activity and plasminogen activator production by hormones, retinoids and the testicular paracrine factor, PModS. Biol. Reprod. 1992; 46:586-594.

108. Moore, A., Findlay, K. and Morris, I.D. In vitro DNA synthesis in Leydig and other interstitial cells of the rat testis. J. Endocrinol. 1992; 134:247-255.

109. Hardy, M.P., Gelber, S.J., Zhou, Z., Penning, T.M., Ricigliano, J.W., Ganjam, V.K., Nonneman, D. and Ewing, L.L. Hormonal control of Leydig cell differentiation. Ann. N.Y. Acad. Sci. 1991; 637:152-163.

110. Bernier, M., Chatelain, P. Mather, J.P. and Saez, J.M. Regulation of gonadotropin receptors, gonadotropin responsiveness and cell multiplication by somatomedin C and insulin in cultured pig Leydig cells. J. Cell. Physiol. 1986; 129:257-263.

110a. Feig, L.A., Bellve, A.R., Horback-Erickson, N. and Klagsbrun, M. Sertoli cells contain a mitogenic polypeptide. Proc. Natl. Acad. Sci. USA 1980; 77:4774-4778.

111. Stubbs, S.C., Hargreave, T.B. and Habib, F.K. Localization and characterization of growth factor receptors on human testicular tissue by biochemical and immuno-histochemical techniques. J. Endocrinol. 1990; 125:485-492.

112. Buch, J.P., Lamb, D.J., Lipschultz, L.J. and Smith, R.G. Partial characterization of a unique growth factor secreted by human Sertoli cells. Fertil. Steril. 1988; 49:658-665.

113. Mallea, L.E., Nachado, A.J., Navaroli, F. and Rommerts, F.F.G. Epidermal growth factor stimulates lactate production and inhibits aromatization in cultured Sertoli cells from immature rats. Int. J. Androl. 1986; 9:201-208.

114. Mullaney, B.P. and Skinner, M.K. Transforming growth factor-α and epidermal growth factor receptor gene expression and action during pubertal development of the seminiferous tubule. Molec. Endcrinol. 1992; 6:2103-2113.

115. Calkins, J.H., Guo, H., Sigel, M.M. and Lin, T. Differential effects of recombinant interleukin-1α and β on Leydig cell function. Biochem. Biophys. Res. Comm. 1990; 167:548-553.

116. Sharpe, R.M., Fraser, H.M., Cooper, I. and Rommerts, F.F. Sertoli-Leydig cell communication via an LHRH-like factor. Nature 1981; 290:785-787.

117. Verhoeven, G. and Cailleau, J.A. A factor in spent media from Sertoli cell-enriched cultures that stimulates steroidogenesis in Leydig cells. Mol. Cell. Endocrinol. 1985; 40:57-68.

118. Cheng, C.Y., Zwain, I., Grima, J. and Bardin, C.W. Structure and functions of Sertoli cell proteins. In: Baccetti, B. ed. Comparative Spermatology 20 Years After. NY: Raven, 1991:617-624.

119. Moore, A. and Morris, I.D. The involvement of insulin-like growth factor-I in local control of steroidogenesis and DNA synthesis of Leydig and non-Leydig cells in the rat testicular interstitium. J. Endocrinol. 1993; 138:107-114.

120. Motro, B., van der Kooy, D., Rossant, J., Reith, A. and Bernstein, A. Contiguous patterns of *c-kit* and *Steel* expression: Analysis of mutations at the *W* and *Sl* loci. Development 1991; 113:1207-1221.
121. Jansz, G.F., Cooke, R.A. and Pomerantz, D.K. Initial characterization of factors from testicular fluid which alter in vitro androgen secretion by normal rat Leydig cells. J. Androl. 1990; 11:131-139.
122. Hedger, M.P., Robertson, D.M., de Kretser, D.M. and Risbridger, G.P. The quantification of steroidogenesis-stimulating activity in testicular interstitial fluid by an in vitro bioassay employing adult rat Leydig cells. Endocrinol. 1990; 127:1967-1977.
123. Martinova, Y.S., Kancheva, L.S., Nikolova, D.B. and Georgiev, V.D. Differential effects of prepuburtal rat Sertoli cell secreted proteins on somatic testicular cells. Molec. and Cell. Endocrinol. 1993; 98:75-79.
124. Boitani, C., Ritzen, E.M. and Parvinen, M. Inhibition of rat Sertoli cell aromatase by factors secreted specifically at spermatogenic stages VII and VIII. Mol. Cell Endocrinol. 1981; 23:11-22.
125. Hall, S.M., Berthelon, M.-C., Avallet, O. and Saez, J.M. Regulation of *c-fos, c-jun, jun-B* and *c-myc* messenger ribonucleic acids by gonadotropin and growth factors in cultured pig Leydig cells. Endocrinol. 1991; 129:1243-1249.
126. Alcivar, A.A., Hake, L.E., Kwon, Y.K. and Hecht, N.B. *junD* mRNA expression differs from *c-jun* and *junB* mRNA expression during male germinal cell differentiation. Mol. Reprod. Devel. 1991; 30:187-193.
127. Lamb, D.J., Tsai, Y.-H., Sternberger, A. and Sanborn, B.M. Sertoli cell nuclear transcriptional activity: Stimulation by follicle stimulating hormone and testosterone in vitro. Endocrinol. 1981; 108:1020-1026.
128. Lamb, D.J., Wagle, J.R., Tsai, Y.-H., Lee, L., Steinberger, A. and Sanborn, B.M. Specificity and nature of the rapid steroid-stimulated increase in Sertoli cell nuclear RNA polymerase activity. J. Steroid Biochem. 1982; 16:653-659.
129. Ailenberg, M., McCAbe, D. and Fritz, I.B. Androgens inhibit plasminogen activator activity secreted by Sertoli cells in culture in a two-chambered assembly. Endocrinol. 1990; 126:1561-1568.
130. Weiner, K.X.B., Pentecost, B.T. and Dias, J.A. Testosterone decreases ornithine decarboxylase messenger RNA levels in primary cultures of rat Sertoli cells. Mol. Endocrinol. 1990; 4:1249-1256.
131. Tindall, D.J. and Means, A.R. Concerning the hormonal regulation of androgen binding protein in rat. Endocrinol. 1976; 99:809-818.
132. Persson, H., Ayer-LeLievre, C., Soder, O., Villar, M.J., Metsis, M., Olson, L., Ritzen, E.M. and Hokfelt, T. Expression of β-nerve growth factor receptor mRNA in Sertoli cells downregulated by testosterone. Science 1990; 247:704-707.
133. Cheng, C.Y. and Bardin, C.W. Indentification of two testosterone-responsive proteins in Sertoli cell-enriched culture medium whose secretion is suppressed by cells of the intact seminiferous tubule. J. Biol. Chem. 1987; 262:12768-12779.
134. Kerr, J.B., Savage, G.N., Millar, M. and Sharpe, R.M. Response of the seminiferous epithelium of the rat testis to withdrawal of androgen: evidence for direct effect upon intercellular spaces associated with Sertoli cell junctional complexes. Cell Tiss. Res. 1993; 274:153-161.

135. Margioris, A.N., Liotta, A., Vaudry, H., Bardin, C.W. and Krieger, D.T. Characterization of immunoreactive proopiomelanocortin-related peptides in rat testis. Endocrinol. 1983; 113:663-671.
136. Fabbri, A., Knox, G., Buzcko, E. and Dufau, M.L. β-endorphin production by the fetal Leydig cell: Regulation and implications for paracrine control of Sertoli cell function. Endocrinol. 1988; 122:749-755.
137. Orth, J.M. FSH-induced Sertoli cell proliferation in the developing rat is modified by β-endorphin produced in the testis. Endocrinol. 1986; 119:1876-1878.
138. Hovatta, O. Effect of androgens and antiandrogens on the development of the myoid cells of the rat seminiferous tubules organ culture. Z. Zellforsch. 1972; 131:299-308.
139. Skinner, M.K. and Fritz, I.B. Testicular peritubular cells secrete a protein under androgen control that modulates Sertoli cell functions. Proc. Natl. Acad. Sci. USA 1985; 82:114-118.
140. Gnessi, L., Emidi, A., Scarpa, S., Palleschi, S., Ragano-Caracciolo, M., Silvestroni, L., Modesti, A. and Spera, G. Platelet-derived growth factor effects on purified testicular myoid cells: Binding, cytosolic Ca^{2+} increase, mitogenic activity and extracellular matrix production enhancement. Endocrinol. 1993; 133:1880-1890.
141. Gnessi, L., Emidi, A., Farini, D., Scarpa, S., Modesti, A., Ciampani, T., Silvestroni, L. and Spera, G. Rat Leydig cells bind platelet-derived growth factor through specific receptors and produce platelet-derived growth factor-like molecules. Endocrinol. 1992; 130:2219-2224.
142. Murono, E.P. and Washburn, A.L. Platelet-derived growth factor inhibits 5α-reductase and Δ5-3β-hydroxysteroid dehydrogenase activities in cultured Leydig cells. Biochem. Biophys. Res. Comm. 1990; 169: 1229-1234.
143. Sun, X.-R., Hedger, M.P. and Risbridger, G.P. The effect of testicular macrophages and interleukin-1 on testosterone production by purified adult rat Leydig cells cultured under in vitro maintenance conditions. Endocrinology 1993; 132:186-192.
144. Risley, M.S., Tan, I.P., Roy, C. and Saez, J.C. Cell-, age-, and stage-dependent distribution of connexin43 gap junctions in testes. J. Cell Sci. 1992; 103:81-96.
145. Raychoudhury, S., Blackshaw, A.W. and Irving, M.G. Hormonal modulation of the interactions of cultured rat testicular Sertoli and peritubular myoid cells. Effects on glycosaminoglycan synthesis. J. Androl. 1993; 14:9-16.
146. Nishiume, Y. and Okabe, M. Mammalian male gametogenesis: Growth, differentiation and maturation of germ cells. Devel. Growth Differ. 1993; 35:479-486.
147. Mizunuma, M., Dohmae, K., Tajima, Y., Koshimizu, U., Watanabe, D. and Nishimune, Y. Loss of sperm in juvenile spermatogonial depletion *jsd* mutant mice is ascribed to a defect of intratubular environment to support germ cell differentiation. J. Cell. Physiol. 1992; 150:188-193.
148. Weinbauer, G.F. and Nieschlag, E. Hormonal control of spermiogenesis. In: deKretser, D. ed. Molecular Biology of the Male Reproductive System. NY: Acad.Press, 1993:99-142.
149. Verhoeven, G. and Cailleau, J. Follicle-stimulating hormone and androgens increase the concentration of androgen receptor in Sertoli cells.

Endocrinol. 1988; 122:1541-1550.
150. Tsutsui, K. Pituitary and gonadal hormone-dependent and -independent induction of follicle-stimulating hormone receptors in the developing testis. Endocrinol. 1991; 128:477-487.
151. Cunningham, G.R. and Huckins, C. Persistence of complete spermatogenesis in the presence of low intratesticular concentrations of testosterone. Endocrinol. 1979; 105:177-186.
152. Rivarola, M.A., Sanchez, P. and Saez, J.M. Stimulation of ribonucleic acid and deoxyribonucleic acid synthesis in spermatogenic cells by their co-culture with Sertoli cells. Endocrinol. 1985; 117:1796-1802.
153. Li, L.Y., Seddon, J.P. Meister, A. and Risley, M.S. Spermatogenic cell-somatic cell interactions are required for maintenance of spermatogenic cell glutathione. Biol. Reprod. 1989; 40:317-331.
154. van der Donk, J.A., de Ruiter-Bootsman, A.L., Ultee-van Gessel, A.M. and Wauben-Penris, P.J. Cell-cell interaction between rat Sertoli cells and mouse germ cells in vitro. Exp. Cell Res. 1986; 164:191-198.
155. Griswold, M.D. Protein secretion of Sertoli cells. Int. Rev. Cytol. 1988; 110:133-156.
156. Pollanen, P., Soder, O. and Parvinen, M. Interleukin-1 α stimulation of spermatogonial proliferation in vivo. Reprod. Fertil. 1989; 1:85-87.
157. Soder, O., Syed, V., Callard, J., Toppari, J., Pollanen, P., Parvinen, M., Froysa, B. and Ritzen, E.M. Production and secretion of an interleukin-1-like factor. Int. J. Androl. 1991; 14:223-231.
158. Parvinen, M., Soder, O., Mali, P., Froysa, B. and Ritzen, E.M. In vitro stimulation of stage-specific DNA synthesis in rat seminiferous tubule segments by interleukin-1α. Endocrinol. 1991; 129:1614-1620.
159. Norton, J.N. and Skinner, M.K. Regulation of Sertoli cell differentiation by the testicular paracrine factor PModS: potential role of immediate-early genes. Mol. Endocrinol. 1992; 6:2018-2023.
160. Dubois, W. and Callard, G.V. Culture of intact Sertoli/germ cell units and isolated Sertoli cells from *Squalus* testis. II. Stimulatory effects of insulin and IGF-I on DNA synthesis in premeiotic stages. J. Exp. Zool. 1993; 267:233-244.
161. Vanelli, B.G., Barni, T., Orlando, C., Natali, A., Serio, M. and Balboni, G.G. Insulin-like growth factor-I IGF-I and IGF-I receptor in human testis: An immunohistochemical study. Fertil. Steril. 1988; 49:666-669.
162. Naz, R.K. and Kumar, R. Transforming growth factor β1 enhances expression of a 50 kDa protein related to 2'-5' oligoadenylate synthetase in human sperm cells. J. Cell. Physiol. 1991; 146:156-163.
163. Tajima, Y., Nishina, Y., Koshimizu, U., Jippo, T., Kitamura, Y. and Nishimune, Y. Effects of hormones, cyclic AMP analogues and growth factors on steel factor SF production in mouse Sertoli cells. J. Reprod. Fertil. 1993; 99:571-575.
164. Mather, J.P., Attie, K.M., Woodruff, T.K., Rice, G.C. and Phillips, D.M. Activin stimulates spermatogonial proliferation in germ-Sertoli cell cocultures from immature rat testis. Endocrinol. 1990; 127:3206-3214.
165. Woodruff, T.K., Borree, J., Attie, K.M., Cox, E.T., Rice, C.G. and Mather, J.P. Stage-specific binding of inhibin and activin to subpopulations of rat germ cells. Endocrinol. 1992; 130:871-881.
166. Bhasin, S., Krummen, L., Morelos, B.S., Kim, W.H., Swerdloff, R.S.,

diZerega, G.S., Ling, N., Esch, F., Shimansaki, S. and Toppari, J. Stage-specific expression of inhibin α and β B subunits during the cycle of the rat seminiferous epithelium. Endocrinol. 1989; 124:987-991.

167. Kaipia, A., Parvinen, M., Shimasaki, S., Ling, N. and Toppari, J. Stage-specific cellular regulation of inhibin α-subunit mRNA expression in the rat seminiferous epithelium. Mol. Cell. Endocrinol. 1991; 82:165-173.

168. Hakovirta, H., Kaipa, A., Soder, O. and Parvinen, M. Effects of activin-A, inhibin-A and transforming growth factor-β1 on stage-specific deoxyribonucleic acid synthesis during rat seminiferous epithelial cycle. Endocrinol. 1993; 133:1664-1668.

169. O'Brien, D.A., Gabel, C.A., Welch, J.E. and Eddy, E.M. Mannose 6-phosphate receptors: Potential mediators of germ cell-Sertoli cell interactions. Ann. N.Y. Acad. Sci. 1991; 637:327-339.

170. O'Brien, D.A., Gabel, C.A. and Eddy, E.M. Mouse Sertoli cells secrete mannose-6-phosphate containing glycoproteins that are endocytosed by spermatogenic cells. Biol. Reprod. 1993; 49:1055-1065.

171. Schaller, J., Glander, H.J. and Detloff, J. Evidence of β1 integrins and fibronectin on spermatogenic cells in human testis. Human Reprod. 1993; 8:1873-1878.

172. Davis, C.M., Papadoupoulos, V., Jia, M., Yamada, Y., Kleinman, H.K. and Dym, M. Identification and partial characterization of laminin binding proteins in immature rat Sertoli cells. Exp. Cell Res. 1991; 193:262-273.

173. Palombi, E., Salanova, M., Tarone, G., Farini, D. and Stefani, M. Distribution of β integrin subunit in rat seminiferous epithelium. Biol. Reprod. 1992; 47:1173-1182.

174. Corlu, A., Gerard, N., Rissel, M., Kercret, H., Kniep, B., Guguen-Guillouzo, C. and Jegou, B. Identification of a plasma membrane protein involved in the cell-cell contact-mediated regulation of the Sertoli cell function. 7th European Workshop on Molecular and Cellular Endocrinology of the Testis; Castle Elmau:Germany.

175. Jegou, B. The Sertoli-germ cell communication network in mammals. Int. Rev. Cytol. 1993; 147:25-96.

176. D'Agostino, A. and Stefanini, M. An intercellular adhesion protein of rat testis: isolation and preliminary characterization. Molec. Reprod. Devel. 1985; 25:172-176.

177. Abdullah, M. and Kierszenbaum, A.L. Identification of rat testis galactosyl receptor using antibodies to liver asialoglycoprotein receptor: purification and localization on the surfaces of spermatogenic cells and sperm. J. Cell Biol. 1989; 108:367-375.

178. Pratt, S.A., Scully, N. and Shur, B.D. Cell surface β1,4 galactosyltransferase on primary spermatocytes facilitates their adhesion to Sertoli cells in vitro. Biol. Reprod. 1993; 49:470-482.

179. Raychoudhury, S.S. and Millette, C.F. Surface-associated glycosyltransferase activities in rat Sertoli cells in vitro. Molec. Reprod. Devel. 1993; 36:195-202.

180. Newton, S.C. and Millette, C.F. Sertoli cell plasma membrane polypeptides involved in spermatogenic cell-Sertoli cell adhesion. Biol. Reprod. 1991; 44 Supp:93-99.

181. Russell, L.D. and Peterson, R.N. Sertoli cell junctions: Morphological and

functional correlates. Int. Rev. Cytol. 1985; 94:177-211.
182. Merchant, H. Rat gonadal and ovarian organogenesis with and without germ cells: An ultrastructural study. Devel. Biol. 1975; 44:1-21.
183. Munsterberg, A. and Lovell-Badge, R. Expression of the mouse anti-Mullerian hormone gene suggests a role in both male and female differentiation. Development 1991; 113:613-624.
184. Jegou, B. Spermatids are regulators of Sertoli cell function. Ann. N.Y. Acad. Sci. 1991; 637:340-353.
185. Jegou, B., Laws, A.O. and deKretser, D.M. Changes in testicular function induced by short-term exposure of the rat testis to heat: Further evidence for interaction of germ cells, Sertoli cells, and Leydig cells. Int. J. Androl. 1984; 7:244-257.
186. Pineau, C., de la Calle, J.F.V., Pinon-Lataillade, G. and Jegou, B. Assessment of testicular function after acute and chronic irradiation: further evidence for an influence of late spermatids on Sertoli cell function in the adult rat. Endocrinol. 1989; 124:2720-2728.
187. Velez de la Calle, J.F., Soufir, J.C.I., Chodorge, F., Boisseau, C., Kercret, H. and Jegou, B. Reproductive effect of the anti-cancer drug procarbazine in male rats at different ages. J. Reprod. Fertil. 1988; 83:51-61.
188. Sharpe, R.M., Millar, M. and McKinnell, C. Relative roles of testosterone and the germ cell complement in determining stage-dependent changes in protein secretion by isolated rat seminiferous tubules. Inter. J. Androl. 1993; 16:71-81.
189. Kamtchouing, P., Pinon-Lataillade, G., Papadopoulous, V., Guillaumin, J.M., Bardos, P., Mass, J., Perreau, C., Drosdowsky, M.A., Hochereau, v.de Reviers, M.T. and Carreau, S. Effect of continuous low-dose γ-irradiation on rat Sertoli cell function . Reprod. Nutr. Dev. 1988; 28:1009-1017.
190. Kangasniemi, M., Kaipia, A., Toppari, J., Mali, P. and Huhtaniemi, I. Cellular regulation of basal and FSH-stimulated cyclic AMP production in irradiated rat testes. Anat. Rec. 1990; 227:32-36.
191. Cheng, C.Y., Grima, J., Stahler, M.S., Lockshin, R.A. and Bardin, C.W. Testins are structurally related Sertoli cell products whose secretion is tighlty coupled to the presence of germ cells. J. Biol. Chem. 1989; 264:21386-21393.
192. Maguire, S.M., Millar, M.R., Sharpe, R.M. and Saunders, P.T.K. Stage-dependent expression of mRNA for cyclic protein 2 during spermatogenesis is modulated by elongate spermatids. Molec. Cell. Endocrinol. 1993; 94:79-88.
193. Zabludoff, S.D., Erickson-Lawrence, M. and Wright, W.W. Sertoli cells proximal convoluted tubules in the kidney and neurons in the brain contain cyclic protein 2. Biol. Reprod. 1990; 43:15-24.
194. Vikho, K.K., Suominen, J.J.O. and Parvinen, M. Cellular regulation of plasminogen activator secretion during spermatogenesis. Biol. Reprod. 1984; 31:383-389.
195. Allenby, G., Foster, P.M.D. and Sharpe, R.M. Evidence that secretion of immunoreactive inhibin by seminiferous tubules from the adult rat testis is regulated by specific germ cell types: correlation between in vivo and in vitro studies. Endocrinol. 1991; 128:467-476.
196. McKinnell, C. and Sharpe, R.M. The role of specific germ cell types in

modulation of the secretion of androgen-regulated proteins by stage VI-VII seminiferous tubules from the adult rat. Mol. Cell. Endocrinol 1992; 83:219-231.

197. Jegou, B., Pineau, C., Velez de la Calle, J.F., Touzalin, A.-M., Bardin, C.W and Cheng, C.Y. Germ cell control of testin production is inverse to that of other Sertoli cell products. Endocrinol. 1993; 132:2557-2562.
198. LeMagueresse, B. and Jegou, B. Paracrine control of immature Sertoli cells by adult germ cells in the rat: an in vitro study. Mol. Cell Endocrinol. 1988; 58:65-72.
199. LeMagueresse, B. and Jegou, B. In vitro effects of germ cells on the secretory pathway of Sertoli cells recovered from rats of different ages. Endocrinol. 1988; 122:1672-1680.
200. Castellon, E., Janecki, A. and Steinberger, A. Age-dependent Sertoli cell responsiveness to germ cells in vitro. Int. J. Androl. 1989; 12:439-450.
201. Trifiro, M., Prior, R.L., Sabbaghian, N., Pinsky, L., Kaufman, M., Nylen, E.G., Belsham, D.D., Greenberg, C.R. and Wrogemann, K. Amber mutation causes a diagnostic MaeI site in the androgen receptor gene of a family with complete androgen insensitivity. Am. J. Med. Genet. 1991; 40:493-499.
202. Stallard, B.J. and Griswold, M.D. Germ cell regulation of Sertoli cell transferrin mRNA levels. Mol. Endocrinol. 1990; 4:393-401.
203. LeMagueresse, B. and Jegou, B. Paracrine control of immature Sertoli cells by adult germ cells in the rat (an in vitro study). Mol. Cell Endocrinol. 1988; 58:65-72.
204. Pineau, C., Sharpe, R.M., Saunders, P.T.K., Gerard, N. and Jegou, F. Regulation of Sertoli cell inhibin production and of inhibin α-subunit mRNA levels by specific germ cell types. Mol. Cell. Endocrinol. 1990; 72:13-22.
205. Schteingart, H.F., Rivarola, M.A. and Cigorraga, S.B. Hormonal and paracrine regulation of γ-glutamyl transpeptidase in rat Sertoli cells. Mol. Cell. Endocrinol. 1989; 67:73-80.
206. Fujisawa, M., Bardin, C.W. and Morris, P.L. A germ cell factor modulates preproenkephalin gene expression in rat Sertoli cells. Mol. Cell. Endocrinol. 1992; 84:79-88.
207. Welsh, M.J. and Ireland, M.E. The second messenger pathway for germ cell mediated stimulation of Sertoli cells. Biochem. Biophys. Res. Comm. 1992; 184:217-227.
208. LeMagueresse, B., LeGac, F., Loir, M. and Jegou, B. Stimulation of the rat Sertoli cell secretory activity in vitro by germ cells and residual bodies. J. Reprod. Fertil. 1986; 77:489-498.
209. Onada, M., Djakiew, D. and Papadopoulos, V. Pachytene spermatocytes regulate the secretion of Sertoli cell proteins which stimulate Leydig cell steroidogenesis. Mol. Cell. Endocrinol. 1991; 77:201-216.
210. LeMagueresse, B. and Jegou, B. Possible involvement of germ cells in the regulation on oestradiol-17β and ABP secretion by immature rat Sertoli cells in vitro studies. Biochem. Biophys. Res. Comm. 1986; 141:861-869.
211. Gerard, N., Syed, V. and Jegou, B. Lipopolysaccharide, latex beads, and residual bodies are potent activators of Sertoli cell interleukin 1-α production. Biochem. Biophys. Res. Comm. 1992; 185:154-161.
212. Syed, V., Gerard, N., Kaipia, A., Bardin, C.W., Parvinen, M. and Jegou,

B. Identification, ontogeny, and regulation of an interleukin-6-like IL-6 factor in the rat testis. Endocrinol. 1993; 132:293-299.

213. Gerard, N. and Jegou, B. In vitro influence of germ cells on Sertoli cell-secreted proteins: a two-dimensional gel electrophoresis analysis. Int. J. Androl. 1993; 16:285-291.

214. Onoda, M. and Djakiew, D. A 29,000 Mr protein derived from round spermatids regulates Sertoli cell secretion. Molec. Cell. Endocrinol. 1993; 93:53-61.

215. Castellon, E., Janecki, A. and Steinberger, A. Influence of germ cells on Sertoli cell secretory activity in direct and indirect co-culture with Sertoli cells from rats of different ages. Mol. Cell Endocrinol. 1989; 64:169-178.

216. Djakiew, D. and Dym, M. Pachytene spermatocyte proteins influence Sertoli cell function. Biol. Reprod. 1988; 39:1193-1205.

217. Onada, M. and Djakiew, D. Pachytene spermatocyte proteins stimulate Sertoli cells grown in bicameral chambers: dose-dependent secretion of ceruloplasmin, sulfated glycoprotein -1, sulfated glycoprotein-2 and transferrin. In Vitro Cell. Dev. Biol. 1991; 27A:215-222.

218. Ireland, M.E. and Welsh, M.J. Germ cell stimulation of Sertoli cell protein phosphorylation. Endocrinol. 1987; 120:1317-1326.

219. Galdieri, M., Ziparo, E., Palombi, F., Russo, M.A. and Stefanini, M. Pure Sertoli cell cultures: a new model for the study of somatic-germ cell interactions. J. Androl. 1981; 2:249-254.

220. LeMagueresse, B., Pineau, C., Guillou, F. and Jegou, B. Influence of germ cells upon transferrin secretion by rat Sertoli cells in vitro. J. Endocrinol. 1988; 118:R13-R16.

221. Pinon-Lataillade, G., Vignier-Martiinez, M.C., Touzalin, A.M., Maas, J. and Jegou, B. Influence of an acute exposure of rat testes to γ-rays on germ cells and on Sertoli and Leydig cell functions. Reprod. Nutr. Dev. 1991; 31:617-629.

222. Onada, M. and Djakiew, D. Modulation of Sertoli cell secretory function by round spermatid proteins. Mol. Cell. Endocrinol. 1990; 73:35-44.

223. Roberts, K.P., Awoniyi, C.A., Santulli, R. and Zirkin, B.R. Regulation of Sertoli cell transferrin and sulfated glycoprotein-2 messenger ribonucleic acid levels during the restoration of spermatogenesis in the adult hypophysectomized rat. Endocrinol. 1991; 129:3417-3423.

224. Baba, T., Hoff, H.B., Nemoto, H., Lee, H., Orth, J., Arai, Y. and Gerton, G.L. Acrogranin, an acrosomal cysteine-rich glycoprotein, is the precursor of the growth-modulating peptides, granulins, and epithelins, and is expressed in somatic as well as male germ cells. Molec. Reprod. Devel. 1993; 34:233-243.

225. Ayer-LeLievre, C.L., Olson, L., Ebendal, T., Hallbook, F. and Persson, H. Nerve growth factor mRNA and protein in the testis and epididymis of mouse and rat. Proc. Natl. Acad. Sci. USA 1988; 85:2628-2632.

226. Lonnerberg, P., Soder, O., Parvinen, M., Ritzen, E.M. and Persson, H. β-Nerve growth factor influences the expression of androgen-binding protein messenger ribonucleic acid in the rat testis. Biol. Reprod. 1992; 47:381-388.

227. Mayerhofer, A., Russell, L.D., Grothe, C., Rudolf, M. and Gratzl, M. Presence and localization of a 30-kDa basic fibroblast growth factor-like protein in rodent testes. Endocrinol. 1991; 129:921-924.

228. Lahr, G., Mayerhofer, A., Seidl, K., Bucher, S., Grothe, C., Knochel, W. and Gratzl, M. Basic fibroblast growth factor bFGF in rodent testis. Presence of bFGF mRNA and of a 30 kDa bFGF protein in pachytene spermatocytes. FEBS Lett. 1992; 302:43-46.
229. Jaillard, C., Chatelain, P.G. and Saez, J.M. In vitro regulation of pig Sertoli cell function: effects of fibroblast growth factor and somatomedin-C. Biol. Reprod. 1987; 37:665-674.
230. Hall, S.H., Joseph, D.R., French, F.S. and Conti, M. Follicle stimulating hormone induces transient expression of the proto-oncogene *c-fos* in primary Sertoli cell cultures. Mol. Endocrinol. 1988; 2:55-61.
231. Smith, E.P., Hall, S.H., Monaco, L., French, F.S., Wilson, E.M. and Conti, M. Regulation of *c-fos* messenger ribonucleic acid by fibroblast growth factor in cultured Sertoli cells. Ann. N.Y. Acad. Sci. 1989; 564:132-139.
232. Smith, E.P., Hall, S.H., Monaco, L. French, F.S., Wilson, E.M. and Conti, M. A rat Sertoli cell factor similar to basic fibroblast growth factor increases *c-fos* messenger ribonucleic acid in cultured Sertoli cells. Mol. Endocrinol. 1989; 3:954-961.
233. Story, M.T., Sasse, J., Kakuska, D., Jacobs, S.D. and Lawson, R.K. A growth factor in bovine and human testes structurally related to basic fibroblast growth factor. J. Urol. 1988; 140:422-427.
234. Han, I.S., Sylvester, S.R., Kim, K.H., Schelling, M.E., Venkateswaran, S., Blanckaert, V.D., McGuiness, M.P. and Griswold, M.D. Basic fibroblast growth factor is a testicular germ cell product which may regulate Sertoli cell function. Molec. Endocrinol. 1993; 7:889-897.
235. Sordoillet, C., Chauvin, M.A., Revol, A., Morera, A.M. and Benamed, M. Fibroblast growth factor is a regulator of testosterone secretion in cultured immature Leydig cells. Mol. Cell. Endocrinol. 1988; 58:283-291.
236. De, S.K., Chen, H.-L., Pace, J.L., Hunt, J.S., Terranova, P.F. and Enders, G.C. Expression of tumor necrosis factor-α in mouse spermatogenic cells. Endocrinol. 1993; 133:389-396.
237. Mealy, K., Robinson, B., Millette, C.F., Majzoub, J. and Wilmore, D.W. The testicular effects of tumor necrosis factor. Ann. Surg. 1990; 211:470-475.
238. Boockfor, F.R. and Schwarz, L.K. Effects of interleukin-6, interleukin-2, and tumor necrosis factor on a transferrin release from Sertoli cells in culture. Endocrinol. 1991; 129:256-262.
239. Srivastava, C.H., Collard, M.W., Rothrock, J.K., Peredo, M.J., Berry, S.A. and Pescovitz, O.H. Germ cell localization of a testicular growth hormone-releasing hormone-like factor. Endocrinol. 1993; 133:83-89.
240. Srivastava, C.H., Breyer, P.R., Rothrock, J.K., Peredo, M.J. and Pescovitz, O.H. A new target for growth hormone releasing-hormone action in rat: The Sertoli cell. Endocrinol. 1993; 133:1478-1481.
241. Radhakrishnan, B., Oke, B.O., Papadopoulos, V., Diaugustine, P. and Suarez-Quian, C.A. Characterization of epidermal growth factor in mouse testis. Endocrinol. 1992; 131:3091-3099.
242. Tuohimaa, P., Blauer, M., Bergmann, M. and Aumuller, G. Change in location and processing of inhibin α-subunit percursors. Endocrinol. 1993; 132:629-633.

CHAPTER 7

Conclusions and Perspectives

Spermatogenesis is an attractive area of investigation for students of differentiation, hormone action, meiosis, morphogenesis, toxicology, and gene expression. However, for certain technical reasons, its full promise has not been realized. Armed with new and precise molecular tools, we can expect rapid advances in the near future. Below are discussed some directions where we may predict the field to progress while we look forward to unanticipated innovations.

IN VITRO SYSTEMS

One of the more frustrating technical problems in the study of spermatogenesis of mammals has been the lack of a suitable culture system in which full differentiation of male germ cells takes place. The advantages of such a system include the ease and accuracy of manipulation of environmental conditions for studying extracellular signaling mechanisms and the ability to express constructs of testis-specific genes to define their controlling elements. Sertoli-spermatogenic cocultures, short term cultures of reaggregated germ cells with Sertoli cells and short term cultures of seminiferous tubules have been put to good use.[1] As expected from the intimate interactions of somatic and germ cells in the testis, culture of isolated germ cells, free of Sertoli cells or their equivalents has been least promising.

Vertebrate culture systems have been reviewed thoroughly in ref. 2. In general, non-mammalian vertebrate and invertebrate organ cultures have proved more fruitful than mammalian cultures, and such systems should probably receive more attention. Some of these cultures have been successful using serum-free chemically defined media.

In cultured silkworm or *Drosophila* testes, primary spermatocytes differentiate to late spermatids.[3-5] In sea urchin organ cultures, spermatogonia or premeiotic spermatocytes differentiate to mid-spermatids.[6]

In newt, isolated spermatocytes develop from early meiotic prophase to round spermatids.[7] In newt organ culture in a chemically defined medium, fragments of testis containing somatic cells and mostly spermatogonia exhibit hormonally controlled differentiation into primary spermatocytes. FSH was shown to be the only hormone needed for this effect.[8]

Frogs and eels have provided two of the best culture systems. In *Xenopus*, a serum-free culture was maintained for 28 days during which time labeled Type B spermatogonia differentiated to spermatids and zygotene spermatocytes became late spermatids.[9] Spermatogonia continued to proliferate throughout.[2] In the eel, HCG or testosterone induce all stages of spermatogenesis within 24 days in a serum-free organ culture suggesting that HCG acts indirectly through Leydig cells.[10,11]

Mouse Type A spermatogonia will differentiate in vitro to pachytene spermatocytes, requiring retinoids and FSH.[12-14] In prepubertal rat testis organ cultures, germ cell maintenance and differentiation require FSH to allow spermatogonia to progress to the pachytene spermatocyte stage, but the proliferation of undifferentiated spermatogonia was independent of FSH.[15] Serum-free media containing growth hormones support spermatogonial division and some signs of differentiation of rat spermatocytes or spermatids (such as histone TH H2B and H1t mRNA synthesis) in Sertoli-germ cell cocultures.[1,16]

Leydig cells can be cultured under conditions which maintain steroidogenic responsiveness.[17] Sertoli function whether alone or in coculture is critically dependent on the type of substrate upon which the cells are cultured.[18,19] Mammalian Sertoli cells in culture may lose their typical morphology unless in the presence of basement lamina matrix[20] or in special chambers on permeable supports.[1,22,23] Then they show polarization, tight junctional attachments, columnar shapes and polarized secretion of transferrin and ABP, but they may still be aberrant with respect to some differentiated functions such as intermediate filament expression and some surface specializations.[18,23] Sertoli cells are often taken from immature animals but may show limited proliferation in culture.[24]

A promising approach to study of germ line cells in culture is the use of SV40 large T antigen to immortalize the cells to establish cell lines. One germ cell, 16 peritubular, and 22 Leydig, 88 Sertoli cell lines were established which exhibited long term culture after transfection with a plasmid containing the large T antigen.[25] These cells produced molecules characteristic of the differentiated states such as alkaline phosphatase, collagen and desmin (peritubular cells), 3-β-hydroxysteroid (Leydig cells), FSH receptors (Sertoli cells) and cytochrome c_T and lactate dehydrogenase C_4 (germ line). The cells can be combined in culture and reconstitute seminiferous cords. Differentiation of germ line cells which corresponds to type B spermatogonia or primary spermatocytes was not seen.

GENE REGULATION

Continued progress will be made in studying the many levels of regulation of male germ line expressed genes. In vitro transcription systems should continue to prove useful in defining *cis*-acting elements and in isolating and testing transcription factors. Gel retardation and DNA footprinting have proved valuable ways of defining regions of interactions. Reporter gene constructs have been problematic because of the lack of permanent cell lines for transfection discussed above. As an alternative transgenic mice have received attention in analyzing these constructs and will continue to be utilized. They have revealed sequences controlling testis-specific, cell-specific and temporal gene expression. In addition, transgenes often alter the spermatogenic process in ways that reveal unanticipated actions of other genes.[26] Construction of "knock-out" mice through homologous recombination is also only beginning to be exploited and should be useful in analyzing those genes not absolutely required for critical developmental events preceding spermatogenesis.

The use of antisense constructs to create dominant negative mutants which eliminate the expression of the endogenous gene and can avoid early lethality has barely been explored in testicular analysis. Antisense constructs have been fruitful in analysis of *wnt-1* during spermatogenesis,[27] and of the effects of PModS acting through c-*fos* in Sertoli cells.[28]

GENETICS

There is no doubt that more mutations affecting spermatogenesis in mice would be desirable. The white spotting and steel mutants *W* and *Sl* mutations have helped to define the *c-kit* and KL interaction critical for stem cell proliferation. The juvenile spermatogonia depletion (*jsd*) mutant causing a defect in intratubular environment is not yet molecularly characterized.[29]

In general, genetics of mice lags behind that of more easily manipulated organisms such as fruit flies and nematodes. Yeast provide a genetic system for investigation of meiosis. Genes isolated and characterized in these organisms provide probes for homologous genes in mammals, as has been exploited for example in isolation of mouse *Hox* genes.

The self-fertilizing hermaphrodite nematode *C. elegans* has permitted the isolation of over 58 genes affecting spermatogenesis (see ref 30). We may expect major additions to our understanding of spermatogenesis from this organism which has contributed so much developmental information on cell lineage, neurogenesis and sex determination.

The relatively small genome size, ease of manipulation, and well-documented genetic system of *Drosophila* makes it an important organism for study of spermatogenesis.[31] Use of enhancer trap techniques in which modified transposable P elements are used to sense active genes when inserted into their regulatory regions is a promising approach to identifying and isolating genes active in the testis.[31] Male sterile mutants blocking spermatogenesis at a variety of stages are available in *Drosophila*.[32] Some block the formation of organelles, others induce heterochronic development of different structures and reveal independent parts of the spermatogenic program. Elegant analyses of the *Drosophila* Y chromosome have been performed. For example, microcloning of Y chromosome regions dissected from lampbrush chromosomes has revealed mostly repetitive non-coding sequences partially homologous to human Y sequences.[33,34]

MOLECULAR CHARACTERIZATION OF STAGE-SPECIFIC MOLECULES

A large number of genes showing spermatid-specific expression has been characterized, as well as many expressed in both spermatocytes and spermatids, but few spermatocyte-specific transcripts like *Rex-1*[35] have been identified. Some laboratories have approached testis-specific gene expression by screening testicular cDNA libraries. One screen turned up 18 DNA clones corresponding to four phases of gene expression: 1) haploid-specific, 2) beginning at pachytene increasing in round and decreasing in elongating spermatids, 3) beginning at leptotene/zygotene increasing in round spermatids and decreasing in elongating spermatids, and 4) expressed in spermatogonia/spermatocytes but not spermatids.[36] In a screen for genes expressed during mouse spermatogenesis, especially during early stages, 26 new genes were isolated that were expressed predominantly in testis.[37] These showed five phases of expression. Three genes were active in spermatogonia and early meiosis. Such screens will yield more molecules expressed in a stage-specific fashion which can then be compared to see if common elements control their transcription, splicing, translation and stability.

SIGNALING AND REGULATORY MECHANISMS

A major shift the in study of hormonal regulation in the testis has occurred with the advent of new molecular tools for the isolation of the genes for peptide hormones and most importantly the genes for hormonal receptors. This work is leading to more refined models of how these molecules are involved in control of spermatogenesis. Much recent effort has gone towards documenting the presence of various growth factors and their receptors in the testis. The results are often difficult to interpret in term of specific effects since many of the factors and receptors are ubiquitous or nearly so, and paracrine and autocrine effects become impossible to elucidate. Advances in characterizing the few apparently testis-specific factors will be welcome.

Understanding of sex determination in vertebrates lags far behind that of *Drosophila* or *C. elegans*. It seems likely that vertebrate mechanisms will differ from those two organisms which use a ratio of X chromosomes to autosomes instead of a Y linked sex determination scheme.[38] In spite of progress in identifying the SRY gene in mammals, we still do not know the targets of its product. There is likely to be a complex genetic pathway involved in mammalian sex determination and it will be more difficult to dissect than from the simpler organisms.

The identification of the *c-kit* oncogene and its ligand offers one of the most successful starting points for understanding mechanisms regulating stem cell proliferation in testis and other tissues. Again, the full pathway involving these molecules awaits further study. Nonetheless it is clear that molecular tools are now powerful enough to encourage us to expect rapid advances in understanding the complex signaling pathways which regulate spermatogenesis.

REFERENCES

1. Tres, L.L., Smith, F.F. and Kierszenbaum, A.L. Spermatogenesis in vitro: methodological advances and cellular functional parameters. In: Negro-Vilar, A. and Perez-Palacios, G. ed. Reproduction, Growth and Development, Serono Symposia. New York: Raven Press, 1991:115-125.
2. Abe, S.-I. Differentiation of spermatogenic cells from vertebrates in vitro. Int. Rev. Cytol. 1987; 109:159-209.
3. Fowler, G.L. In vitro cell differentiation in the testes of *Drosophila hydei*. Cell Differ. 1973; 2:33-42.

4. Liebrich, W. In vitro spermatogenesis in *Drosophila.* I. Development of isolated spermatocytes from wild-type *D. hydei.* Cell Tiss. Res. 1981; 220:251-262.
5. Kambysellis, M.P. and Williams, C.M. Spermatogenesis in cultured testes of the *Cynthia* silkworm: effects of ecdysone and of prothoracic glands. Science 1972; 175:769-770.
6. Poccia, D. In vitro differentiation of male germ line cells from sea urchin testis. J. Exp. Zool. 1988; 246:57-64.
7. Nishikawa, A. and Abe, S.I. Progression through all stages of meiosis from early prophase of newt primary spermatocytes in vitro. Devel. Growth Differ. 1983; 25:323-331.
8. Ji, Z.-S., Kubokawa, K., Ishii, S. and Abe, S.-I. Differentiation of secondary spermatogonia to primary spermatocytes by mammalian follicle-stimulating hormone in organ culture of testes fragments from the newt, *Cynops pyrrhogaster.* Devel. Growth Differ. 1992; 34:649-660.
9. Risley, M.S., Miller, A. and Bumcrot, D.A. In vitro maintenance of spermatogenesis in *Xenopus laevis* testis explants cultured in serum-free media. Biol. Reprod. 1987; 36:985-997.
10. Miura, T., Yamauchi, K., Takahashi, H. and Nagahama, Y. Human chorionic gonadotropin induces all stages of spermatogenesis in vitro in the male Japanese eel *Anguilla japonica.* Devel. Biol. 1991; 146:258-262.
11. Miura, T., Yamauchi, K., Takahasi, H. and Nagahama, Y. Hormonal induction of all stages of spermatogenesis in vitro in the male Japanese eel *Anguilla japonica.* Proc. Natl. Acad. Sci USA 1991; 88:5774-5778.
12. Steinberger, A. In vitro techniques for the study of spermatogenesis. Meth. Enzymol. 1975; 39:283-296.
13. Haneji, T., Maekawa, M. and Nishimune, Y. In vitro differentiation of type A spermatogonia from mouse cryptorchid testes in serum-free media. Biol. Reprod. 1983; 28:1217-1223.
14. Haneji, T., Maekawa, M. and Nishimune, Y. Vitamin A and follicle-stimulating hormone synergistically induce differentiation of type A spermatogonia in adult mouse cryptorchid testes in vitro. Endocrinol. 1984; 114:801-805.
15. Boitani, C., Politi, M.G. and Menna, T. Spermatogonial cell proliferation in organ culture of immature rat testis. Biol. Reprod. 1993; 48:761-767.
16. Tres, L.L., Smith, E.P., van Wyk, J.J. and Kierszenbaum, A.L. Immunoreactive sites and accumulation of somatomedin-C in rat Sertoli-spermatogenic cell co-cultures. Exp. Cell Res. 1986; 162:33-50.
17. Sun, X.-R., Hedger, M.P. and Risbridger, G.P. The effect of testicular macrophages and interleukin-1 on testosterone production by purified adult rat Leydig cells cultured under in vitro maintenance conditions. Endocrinology 1993; 132:186-192.
18. Dym, M. and Papadopoulos, V. Dual compartment bicameral culture: Role of basement membrane in epithelial differentiation. Cell Biol. Toxicol. 1992; 8:55-59.
19. Tres, L.L., Cahn, F. and Kierszenbaum, A.L. Permeable collagen-glycosaminoglycan cross-linked copolymers for the study of biological responses of cocultured Sertoli and spermatogenic cells. Cell Biol. Toxicol. 1992; 8:61-71.
20. Byers, S.W., Hadley, M.A., Djakiew, D. and Dym, M. Growth and characterization of polarized monolayers of epididymal epithelial cells and Sertoli cells grown in dual environment culture chambers. J. Androl. 1986; 7:59-68.

21. Djakiew, D. and Dym, M. Pachytene spermatocyte proteins influence Sertoli cell function. Biol. Reprod. 1988; 39:1193-1205.
22. Janecki, A. and Steinberger, A. Polarized Sertoli cell functions in a new two-compartment culture system. J. Androl. 1986; 7:69-71.
23. Flechon, J.E. and Guillou, F. Differentiation of rat Sertoli cells in vitro. In: Baccetti, B. ed. Comparative Spermatology 20 Years After. NY: Raven, 1991:647-654.
24. Jegou,B. The Sertoli cell in vivo and in vitro. Cell Biol. Toxicol. 1992; 8:49-54.
25. Hofmann, M.-C., Narisawa, S., Hess, R.A. and Millan, J.L. Immortalization of germ cells and somatic testicular cells using the SV40 large T antigen. Exp. Cell Res. 1992; 201:417-435.
26. Eddy, E.M., Welch, J.E. and O'Brien, D.A. Gene expression during spermatogenesis. In: D. deKretser, ed. Molecular Biology of the Male Reproductive System. NY: Acad.Press, 1993:181-232.
27. Erickson, R.P., Lai, L.-W. and Grimes, J. Creating a conditional mutation of *Wnt-1* by antisense transgenesis provides evidence that *Wnt-1* is not essential for spermatogenesis. Devel. Genetics 1993; 14:274-281.
28. Norton, J.N. and Skinner, M.K. Regulation of Sertoli cell differentiation by the testicular paracrine factor PModS: Potential role of immediate-early genes. Molec. Endocrinol. 1992; 6:2018-2026.
29. Nishiume, Y. and Okabe, M.,. Mammalian male gametogenesis: Growth, differentiation and maturation of germ cells. Devel. Growth Differ. 1993; 35:479-486.
30. L'Hernault, S.W., Benian, G.M. and Emmons, R.B. Genetic and molecular characterization of the *Caenorhabditis elegans* spermatogenesis-defective gene *spe-17.* Genetics 1993; 134:769-780.
31. Hennig,W. *Drosophila* spermatogenesis as a model system. Andrologia 1992; 24:21-26.
32. Hackstein, J.H.P. Spermatogenesis in *Drosophila.* A genetic approach to cellular and subcellular differentiation. Eur. J. Cell Biol. 1991; 56:151-169.
33. Vogt, P., Keil, R., Kohler, M., Lengauer, C., Lewe, D. and Lewe, G. Selection of DNA sequences from interval-6 of the human Y-chromosome with homology to a Y-chromosomal fertility gene sequence of *Drosophila hydei.* Hum. Genet. 1991; 86:341-349.
34. Hennig,W. The Y chromosome of *Drosophila.* In: Adolph, K.W., ed. Chromosomes: Eukaryotic, Prokaryotic and Viral. Boca Raton, FL: CRC Press, 1990:213-238.
35. Rogers, M.B., Hosler, B.A. and Gudas, L.J. Specific expression of a retinoic acid-regulated, zinc finger gene, *Rex-1,* in preimplantation embryos, trophoblast and spermatocytes. Development 1991; 113:815-824.
36. Thomas, K.H., Wilkie, T.M., Tomashefsky, P., Bellve, A.R. and Simon, M.I. Differential gene expression during mouse spermatogenesis. Biol. Reprod. 1989; 41:729-739.
37. Starborg, M., Brundell, E. and Hoog, C. Analysis of the expression of a large number of novel genes isolated from mouse prepubertal testis. Molec. Reprod. Devel. 1992; 33:243-251.
38. McKeown, M. and Madigan, S.J. Sex determination and differentiation in invertebrates: *Drosophila* and *Caenorhabditis elegans.* Curr. Opin. Cell Biol. 1992; 4:948-954.

INDEX

Page numbers in italics denote figures (f) and tables (t).

E

F

G

H

I

J